Springer Series in Reliability Engineering

T0137844

Series Editor

Professor Hoang Pham
Department of Industrial Engineering
Rutgers
The State University of New Jersey
96 Frelinghuysen Road
Piscataway, NJ 08854-8018
USA

Other titles in this series

Toshio Nakagawa

Shock and Damage Models in Reliability Theory

 Springer

Toshio Nakagawa, PhD
Department of Marketing and Information Systems
Aichi Institute of Technology
1247 Yachigusa, Yakusa-cho
Toyota 470-0392
Japan

British Library Cataloguing in Publication Data
Nakagawa, Toshio, 1942-
   Shock and damage models in reliability theory. - (Springer
   series in reliability engineering)
   1. Reliability (Engineering) - Mathematical models
   I. Title
   620'.00452'015118

Springer Series in Reliability Engineering series ISSN 1614-7839
ISBN 978-1-84996-601-6   e-ISBN 1-84628-442-2   Printed on acid-free paper
e-ISBN 978-1-84628-442-7

9 8 7 6 5 4 3 2 1

Springer Science+Business Media
springer.com

Preface

Most engineering systems suffer some deterioration with time from *wear, fatigue,* and *damage,* and ultimately fail when their strength exceeds a critical level. Failure mechanisms by which the causes of failures are brought about are physical processes. The types of failure causes, how to proceed to failure by which causes, and the consequences of failures have been physically studied. This has been developed in fracture mechanics and mechanics of materials and has applied to such components and systems. On the other hand, failure mechanisms are in probabilistic and stochastic motions. Such behaviors are mathematically observed and analyzed in the study of stochastic processes.

My purpose in writing this book is to build a *bridge between theory and practice* and to introduce the reliability engineer to some damage models. Failures of units are generally classified into two failure modes: Catastrophic failure in which units fail suddenly and degradation failure in which units deteriorate gradually with time. The former failures often occur in electric parts. The latter failures mainly occur in machinery. Such reliability models are called *shock* or *damage models* and can be analyzed, using the techniques of stochastic processes.

There exist a large number of damage models that form reliability models mechanically and stochastically in the real world. Reliability quantities of these models have been theoretically obtained. However, there is not any special book written on these fields except the book [2]. Their case studies for reliability are very fews because the analysis might be too difficult theoretically to apply them to practical models. When and how maintenance policies for damage models are made are important.

I have just published the monograph *Maintenance Theory of Reliability* [1] that summarizes maintenance policies for system reliability models. However, it does not deal with any damage model. This book is based mainly on the research results studied by the author and my colleagues from classical ones to new topics. It deals primarily with shock and damage models, their reliability properties, and maintenance policies. The reliability measures of such models can be calculated by using renewal and cumulative processes. Optimum maintenance policies are theoretically discussed by using the results of [1]. Furthermore, these models can be applied to actual models practically, using these results.

This book is composed of ten chapters. Chapter 1 gives some examples of damage models and is devoted to explaining elementary stochastic processes and shock processes needed for understanding their models. Chapter 2 is mainly devoted to cumulative damage models that fail subject to shocks. Standard models in which a unit fails when its total damage exceeds a failure level are explained, and their modified models are proposed. Some reliability quantities of such models are analytically derived, using the techniques of stochastic processes. Chapter 3 summarizes replacement policies and some modified policies. Chapter 4 is devoted to a parallel system whose units fail subject to shocks and a two-unit system whose units fail by interaction with induced failure and shock damage. Chapters 5 and 6 are devoted to replacement and preventive maintenance policies in which the total damage is investigated only at periodic times. Chapter 7 considers imperfect preventive maintenance policies in which the preventive maintenance is done at sequential times and reduces the total damage. In Chapters 4–7, optimum policies that minimize the expected cost are analytically discussed. Chapters 8 and 9 take up the garbage collection of a computer system and the backup scheme of a database system as typical practical examples of damage models. Chapter 10 is devoted to reviewing briefly similar related models presented in other fields such as shot noise, insurance, and stochastic duels.

This book gives a detailed introduction to damage models and their maintenance policies, and provides the current status and further studies in these fields. It will be helpful for mechanical engineers and managers engaged in reliability work. Furthermore, sufficient references leading to further studies are cited at the end of the book. This book will serve as a textbook and reference book for graduate students and researchers in reliability and mechanics.

I wish to thank Professor Shunji Osaki for Chapter 2, Dr. Kodo Ito for Chapters 1 and 3, Professor Masaaki Kijima for Chapters 4 and 7, Professor Kazumi Yasui for Chapter 6, Dr. Takashi Satow for Chapter 8, and Professors Cun Hua Qian and Shouji Nakamura who are co-workers of our research papers for Chapter 9. I wish to express my special thanks to Professor Fumio Ohi for his careful reviews of this book, and to Dr. Satoshi Mizutani and my daughter Yorika for their support in writing and typing this book. Finally, I would like to express my sincere appreciation to Professor Hoang Pham, Rutgers University, and editor, Anthony Doyle, Springer-Verlag, London, for providing the opportunity for me to write this book.

Toyota, Japan

Toshio Nakagawa
June 2006

Contents

1

Introduction

The number of aged fossil-fired power plants is increasing in Japan. For example, about one-third of such plants are currently operating at from 150 thousand to 200 thousand hours (from 17 to 23 years), and about a quarter of them are above 200 thousand hours. Furthermore, public infrastructures in advanced nations will become obsolete in the near future [3]. A deliberate maintenance plan is indispensable to operate power and chemical plants without serious trouble.

The importance of maintenance for aged plants is much higher than that for new ones because the probability of the occurrence of severe events increases and new failure phenomena might appear according to the degradation of plants. Actual lifetimes of plant components such as steam and gas turbines, boilers, pipes, and valves, are almost different from predicted ones because they are affected by various factors such as material quality and operating conditions [4, 5]. Therefore, maintenance plans have to be reestablished at appropriate times during the operating lives of these components.

The simplest damage model is the stress-strength model where a component fails when its strength has been below a critical stress level [6]. If the fatigue subject to varying stress can be estimated, Miner's rate can be applied directly, using an $S-N$ curve [7,8]. This is utilized widely for predicting lifetimes of various kinds of mechanical productions by modifying Miner's rule [9].

The progress of physical damage to assess the life of components precisely would be made previously and accurately. For example, the progress of low alloy steel that is used for high temperature and pressure components of a thermal power plant, is observed with a microscope as follows: During the first half of the life, changes in its microstructure appear in the welded heat-affected area. During the latter half, the number of voids that are small cavities at boundaries between crystalline grains increases, and their coalescence results in the growth of a crack. Recently, such damage assessment and life estimation are actively performed by utilizing a digital microscope, a computer image processor, and software [10].

Failures of units or systems such as parts, equipment, components, devices, materials, structures, and machines are generally classified into two failure modes: Catastrophic failure in which units fail by some sudden shock and degradation failure in which units fail by physical deterioration suffered from some damage. In the latter case, units fail when the total damage due to shocks has exceeded a critical failure level. This is called a *cumulative damage model* or *shock model* with additive damage and can be described theoretically by a cumulative process [11] in stochastic processes.

We can apply such damage models to actual units that are working in industry, service, information, and computers, and show typical examples that are familiar.

(1) A vehicle axle fails when the depth of a crack has exceeded a critical level. In actual situations, a train axle is replaced at the distance traveled or the number of revolutions [12]. A tire on an automobile is a similar example [2, 13].

(2) A battery supplies electric power that was stored by chemical change. It is weakened by use and becomes useless at the end of chemical change [14]. This corresponds to the damage model by replacing shock with use and damage with oxidation or deoxidation.

(3) The strength of a fibrous carbon composite is essentially determined by the strength of fibers. When a composite specimen is placed under tensile stress, the fibers themselves may break within the material. Such materials are broken based on cumulative damage [15, 16].

(4) Garbage collection in a database system is a simple method to reclaim the location of active data because updating procedures reduce storage areas and worsen processing efficiency. To use storage areas effectively and to improve processing efficiently, garbage collections are done at suitable times. Such a garbage collection model corresponds to the damage model by replacing shock with update and damage with garbage. Some garbage collection models will be discussed analytically in Chapter 8.

(5) The data in a computer system are frequently updated by adding or deleting them, and are stored in secondary media. However, data files are sometimes broken by several errors due to noises, human errors, and hardware faults. The most dependable method to ensure the safety of data takes their backup copies at appropriate times. This corresponds to the damage model by replacing shock with update and damage with dumped files, and will be discussed analytically in Chapter 9.

Furthermore, damage models were applied to crack growth models [2, 17–20] and to welded joints [21], floating structures [22], reinforced concrete structures [23], and plastic automotive components [24]. Such stochastic models of fatigue damage of materials were described in detail [25, 26]. Failure mechanisms of damage models in engineering systems were summarized [27].

We consider a typical cumulative damage model in which shocks occur in random times and the damage incurred such as fatigue, wear, crack growth, creep, and dielectric breakdown is additive. The general concept of such processes was theoretically based on [28, 29]. Several contributions to stochastic damage models or compound Poisson processes were made at the beginning by several authors: The first model, where shocks occur in a Poisson process and the amount of damage due to each shock has a gamma distribution, was considered in detail [30]. Much of the earlier research were reviewed [11]. Furthermore, the various properties of failure distributions when shocks occur in a Poisson process were extensively investigated [31–33]. On the other hand, cumulative wear increases continuously with time and is represented as a specified function of a stochastic process [34–39]. This was formulated and analyzed by using the idea of a finite Markov chain [2]. This is also called a wear process.

We have to pay attention only to the essential laws governing objective models of reliability study, and grasp damage processes, and try to formulate them simply, avoiding small points. In other words, it would be necessary to form stochastic models of causing and making up damage that outline the observational and theoretical features of complex phenomena.

Most of the contents of this book are based on the original work of our research group and some new results are added. Stochastic and shock processes needed for learning damage models are summarized briefly in Chapter 1. These results are introduced without detailed explanations and proofs.

Chapter 2 summarizes only the known results of cumulative damage models and their modified models based on [11, 33, 40], that could be applied to maintenance policies discussed in the following chapters. Next, we survey briefly the damage model whose total amount increases with time [37, 39, 41].

Suppose that a unit subject to shocks is replaced with a new one at failure or undergoes corrective maintenance after failure. However, such maintenance after failure may be done at great cost and take a long time. The most important problem of maintenance policies is to determine in advance when and how to do better maintenances before failure. From these points of view, a wide variety of uses for maintenance policies are effectively summarized and their optimum policies are fully discussed [1].

The optimum policies for a cumulative damage model where a unit is replaced before failure at a threshold level of damage [42–45] or at a planned time [46–50] were derived. In Chapter 3, we consider three replacement policies for a cumulative damage model in which a unit is replaced before failure at a planned time, at a shock number, or at a managerial damage level [51]. Optimum replacement policies that minimize the expected cost rates are discussed analytically. Furthermore, extended replacement models in which a unit is replaced at the first shock over a planned time and shock number are proposed.

Most systems are composed of multicomponent systems. However, in general, it would be very difficult to analyze the damage models of such systems theoretically. We consider a system with n different units each of which receives damage due to shock and derive the failure distribution of the system in **(4)** of Section 2.5. Furthermore, in Chapter 4, we take up a parallel system in a random environment [52, 53] and consider two models of a two-unit system with failure interactions [54]. Optimum number of units for a parallel system and the number of failures for an interaction model that minimize the expected cost rates are derived.

We should do only some minimal maintenance at each failure in large and complex systems. This is called periodic replacement with minimal repair at failures in Chapter 4 of [1]. In Chapter 5, a unit fails with a certain probability for the total damage due to shocks and undergoes minimal repair. Then, a unit is replaced at a planned time, at a shock number, or at a managerial damage level. In this case, optimum replacement policies that minimize the expected cost rates are discussed analytically [55].

Most operating units are repaired when they have failed. However, it may require much time and high cost to repair a failed unit. The respective maintenance after failure and before failure is called corrective maintenance (CM) and preventive maintenance (PM). This becomes the same as the replacement model theoretically by taking CM and PM as the replacement after failure and before failure, respectively, and the repair time as the time required for replacement.

In Chapter 6, we take up the PM policy in which the test to investigate some characteristics of a unit is planned at periodic times and the PM is done at a planned time when the total damage or shock number has exceeded a managerial level or number [56]. Several modified models are considered and their expected cost rates are derived. Furthermore, in Chapter 7, we apply the imperfect PM model to a cumulative damage model in which the total damage decreases at each PM. An optimum sequential PM policy in which a unit has to be operating over a finite interval and is replaced at a specified PM number is computed numerically [57].

In Chapters 8 and 9, we apply the cumulative damage model to the garbage collection policy [58] and the backup policy for a computer system [59] as typical examples, respectively. Optimum policies that the garbage collection is done at a planned time or at an update number are derived. Three schemes as recovery techniques are introduced, and optimum backup times are discussed analytically and compared numerically.

Such phenomena have been observed frequently in probability fields. Finally, we present compactly in Chapter 10 that the damage model can be applied to related fields such as other reliability models, insurance, shot noise, and stochastic duels. Several quantities of such models are similarly derived, using the techniques of shock and damage models.

1.1 Renewal Processes

In this section, we briefly introduce some basic properties of renewal processes for reliability systems based on the books [11, 60, 61]. For more detailed results

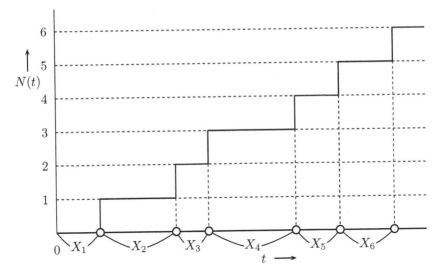

Fig. 1.1. Total number of failed units over time axis

and applications of stochastic processes, we refer readers to the books [62, 63]. Consider a one-unit system with repair or replacement whose time is negligible, *i.e.*, a new unit starts to operate at time 0 and is repaired or replaced when it fails, where the time for repair or replacement is negligible. When the repair or replacement is completed, the unit begins to operate again. If the unit is like new after repair or replacement, then the system forms a *renewal process*. This arises from the study of *self-renewing* aggregates [11] and plays an important role in the analysis of probability models with sums of independent nonnegative random variables. Figure 1.1 is a sample graph that presents the total number $N(t)$ of failed units during a time interval $[0, t]$. Some plots of number of failures versus time for repairable systems were illustrated [64]. In that case, the counting process $\{N(t); t \geq 0\}$ is called a renewal process. In particular, when the unit fails exponentially, *i.e.*, the times between failures are independent and identically distributed exponentially, a renewal process becomes a *Poisson process*. A Poisson process is dealt with frequently as a special case of a renewal process. On the other hand, if the unit after repair has the same age as that before repair, then the counting process $\{N(t); t \geq 0\}$ is called a *nonhomogeneous Poisson process*. This corresponds to the unit that undergoes minimal repair at each failure.

(1) Renewal Process

Consider a sequence of independent and nonnegative random variables $\{X_1, X_2, \cdots\}$, in which $\Pr\{X_j = 0\} < 1$ for all j because of avoiding the triviality.

Suppose that X_j $(j = 1, 2, \cdots)$ have an identical distribution $F(t)$ with finite mean μ_1 and $F(0) \equiv 0$.

Letting $S_n \equiv \sum_{j=1}^{n} X_j$ $(n = 1, 2, \cdots)$ and $S_0 \equiv 0$, we define $N(t) \equiv \max_n \{S_n \le t\}$ that represents the number of renewals in $[0, t]$. Renewal theory is mainly devoted to the investigation into the probabilistic properties of a discrete random variable $N(t)$.

Denote

$$F^{(0)}(t) \equiv \begin{cases} 1 & \text{for } t \ge 0 \\ 0 & \text{for } t < 0 \end{cases} \qquad F^{(n)}(t) \equiv \int_0^t F^{(n-1)}(t-u)\, dF(u) \quad (n = 1, 2, \cdots),$$

i.e., $F^{(n)}(t)$ represents the distribution of $\sum_{j=1}^{n} X_j$. Evidently,

$$\Pr\{N(t) = n\} = \Pr\{S_n \le t \text{ and } S_{n+1} > t\}$$
$$= F^{(n)}(t) - F^{(n+1)}(t) \qquad (n = 0, 1, 2, \cdots). \tag{1.1}$$

We define the expected number of renewals in $[0, t]$ as $M(t) \equiv E\{N(t)\}$, that is called a *renewal function*, and $m(t) \equiv dM(t)/dt$, that is called a *renewal density*. From (1.1),

$$M(t) = \sum_{n=1}^{\infty} n \Pr\{N(t) = n\} = \sum_{n=1}^{\infty} F^{(n)}(t). \tag{1.2}$$

It is fairly easy to show that $M(t)$ is finite for all $t \ge 0$ because $\Pr\{X_j = 0\} < 1$. Furthermore, from the notation of convolution,

$$M(t) = F(t) + \sum_{n=1}^{\infty} \int_0^t F^{(n)}(t - u)\, dF(u)$$
$$= \int_0^t [1 + M(t - u)]\, dF(u), \tag{1.3}$$

that is called a *renewal equation*. When $F(t)$ has a density function $f(t)$ and $f^{(n)}(t) \equiv dF^{(n)}(t)/dt$ $(n = 1, 2, \ldots)$, $m(t) = \sum_{n=1}^{\infty} f^{(n)}(t)$ and differentiation of (1.3) with respect to t implies

$$m(t) = f(t) + \int_0^t m(t - u) f(u)\, du. \tag{1.4}$$

The renewal-type equation such as (1.3) and (1.4) appears frequently in the analysis of stochastic reliability models because most systems are renewed after maintenance. The Laplace–Stieltjes (LS) transform of $M(t)$ is given by

$$M^*(s) \equiv \int_0^\infty e^{-st}\, dM(t) = \frac{F^*(s)}{1 - F^*(s)}, \tag{1.5}$$

where, in general, $\varphi^*(s)$ is the LS transform of $\varphi(t)$, *i.e.*, $\varphi^*(s) \equiv \int_0^\infty e^{-st} d\varphi(t)$ for $s > 0$ and $\int_0^\infty e^{-st} dF^{(n)}(t) = [F^*(s)]^n$ $(n = 0, 1, 2, \dots)$. Thus, $M(t)$ and $F(t)$ determine one another because the LS transform also determines the function uniquely.

The second moment of $N(t)$ is [61, p. 89], because $n^2 = 2 \sum_{i=1}^n i - n$,

$$
\begin{aligned}
E\left\{N(t)^2\right\} &= \sum_{n=1}^\infty n^2 \Pr\left\{N(t) = n\right\} \\
&= 2 \sum_{n=1}^\infty n \Pr\left\{N(t) \geq n\right\} - M(t) \\
&= 2 \sum_{n=1}^\infty n \Pr\left\{S_n \leq t\right\} - M(t) \\
&= 2 \sum_{n=1}^\infty n F^{(n)}(t) - M(t).
\end{aligned}
\tag{1.6}
$$

Forming the LS transforms on both sides above,

$$
\begin{aligned}
\int_0^\infty e^{-st}\, dE\left\{N(t)^2\right\} &= 2 \sum_{n=1}^\infty n [F^*(s)]^n - M^*(s) \\
&= 2 \left[\frac{F^*(s)}{1 - F^*(s)}\right]^2 + \frac{F^*(s)}{1 - F^*(s)} \\
&= 2[M^*(s)]^2 + M^*(s).
\end{aligned}
\tag{1.7}
$$

Inverting (1.7),

$$
E\left\{N(t)^2\right\} = 2M(t) * M(t) + M(t),
\tag{1.8}
$$

and hence,

$$
V\left\{N(t)\right\} = 2M(t) * M(t) + M(t) - [M(t)]^2,
\tag{1.9}
$$

where the asterisk denotes the pairwise Stieltjes convolution, *i.e.*, $a(t) * b(t) \equiv \int_0^t b(t-u) da(u)$.

We summarize some important limiting theorems and results of renewal theory for future reference [11, 60, 61].

Theorem 1.1.

(i)

$$
\frac{M(t)}{t} \longrightarrow \frac{1}{\mu_1}, \qquad \text{as } t \to \infty.
\tag{1.10}
$$

(ii)

$$
\frac{V\left\{N(t)\right\}}{t} \longrightarrow \frac{\sigma^2}{\mu_1^3}, \qquad \text{as } t \to \infty.
\tag{1.11}
$$

Theorem 1.2. If $\mu_2 \equiv \int_0^\infty t^2 dF(t) < \infty$ and $\sigma^2 \equiv \mu_2 - \mu_1^2$,

$$M(t) = \frac{t}{\mu_1} + \left(\frac{\sigma^2}{2\mu_1^2} - \frac{1}{2}\right) + o(1), \qquad \text{ast} \to \infty, \qquad (1.12)$$

and if $\mu_3 \equiv \int_0^\infty t^3 dF(t) < \infty$,

$$V\{N(t)\} = \frac{\sigma^2 t}{\mu_1^3} + \left(\frac{5\sigma^4}{4\mu_1^4} + \frac{2\sigma^2}{\mu_1^2} + \frac{3}{4} - \frac{2\mu_3}{3\mu_1^3}\right) + o(1), \qquad \text{as } t \to \infty, \quad (1.13)$$

where the function $f(h)$ is said to be $o(h)$ if $\lim_{h\to 0} f(h)/h = 0$.
This is proved as follows: Expanding $F^*(s)$ with respect to s,

$$F^*(s) = 1 - \mu_1 s + \frac{1}{2}(\sigma^2 + \mu_1^2)s^2 - \frac{1}{3!}\mu_3 s^3 + o(s^3). \qquad (1.14)$$

Substituting (1.14) in (1.5) and arranging them,

$$M^*(s) = \frac{1}{s\mu_1} + \left(\frac{\sigma^2 - \mu_1^2}{2\mu_1^2}\right) + o(1), \qquad (1.15)$$

$$[M^*(s)]^2 = \frac{1}{s^2\mu_1^2} + \frac{1}{s}\left(\frac{\sigma^2 - \mu_1^2}{\mu_1^3}\right) + \left(\frac{3\sigma^4}{4\mu_1^4} + \frac{\sigma^2}{2\mu_1^2} + \frac{3}{4} - \frac{\mu_3}{3\mu_1^3}\right) + o(1). \qquad (1.16)$$

Inverting (1.15), and substituting (1.16) in (1.7) and inverting it, we have the results of Theorem 1.2 from (1.9).

From this theorem, $M(t)$ and $m(t)$ are approximately given by

$$M(t) \approx \frac{t}{\mu_1} + \frac{\sigma^2}{2\mu_1^2} - \frac{1}{2}, \qquad m(t) \approx \frac{1}{\mu_1}, \qquad (1.17)$$

and

$$V\{N(t)\} \approx \frac{\sigma^2 t}{\mu_1^3} + \frac{5\sigma^4}{4\mu_1^4} + \frac{2\sigma^2}{\mu_1^2} + \frac{3}{4} - \frac{2\mu_3}{3\mu_1^3} \qquad (1.18)$$

for large t. Furthermore, if $\sigma \ll \mu_1$, then

$$M(t) \approx \frac{t}{\mu_1} - \frac{1}{2}. \qquad (1.19)$$

When $F(t)$ has a density function $f(t)$, the failure or hazard rate is defined as $h(t) \equiv f(t)/\overline{F}(t)$, where $\overline{F}(t) \equiv 1 - F(t)$. If the failure rate $h(t)$ is increasing, then F is IFR, that means increasing failure rate.

Theorem 1.3. When F is IFR [65],

$$\frac{t}{\mu_1} - 1 \le \frac{t}{\int_0^t \overline{F}(u)\,du} - 1 \le M(t) \le \frac{tF(t)}{\int_0^t \overline{F}(u)\,du} \le \frac{t}{\mu_1}. \qquad (1.20)$$

Using the asymptotic properties in (1.17) and (1.18) and applying them to the usual central limit theorem, we have the central limit theorem for a renewal process.

Theorem 1.4.

$$\lim_{t \to \infty} \Pr\left\{ \frac{N(t) - t/\mu_1}{\sqrt{\sigma^2 t/\mu_1^3}} \le x \right\} = \frac{1}{\sqrt{2\pi}} \int_{-\infty}^{x} e^{-u^2/2} \, du, \qquad (1.21)$$

i.e., $N(t)$ is asymptotically normally distributed with mean t/μ_1 and variance $\sigma^2 t/\mu_1^3$ for large t.

(2) Poisson Process

When $F(t) = \Pr\{X_j \le t\} = 1 - e^{-\lambda t}$ $(j = 1, 2, \cdots)$ for $\lambda > 0$, the counting process $\{N(t); t \ge 0\}$ is called a *Poisson process* with rate λ. In this case,

$$F^{(n)}(t) = \Pr\{S_n \le t\} = \sum_{j=n}^{\infty} \frac{(\lambda t)^j}{j!} e^{-\lambda t} \qquad (n = 0, 1, 2, \cdots), \qquad (1.22)$$

$$f^{(n)}(t) \equiv \frac{dF^{(n)}(t)}{dt} = \frac{\lambda(\lambda t)^{n-1}}{(n-1)!} e^{-\lambda t} \qquad (n = 1, 2, \cdots), \qquad (1.23)$$

that is a gamma or Erlang distribution with rate λ. From (1.1), (1.2), (1.9), and (1.22), we easily have the following results:

$$\Pr\{N(t) = n\} = \frac{(\lambda t)^n}{n!} e^{-\lambda t} \qquad (n = 0, 1, 2, \cdots), \qquad (1.24)$$

i.e., $N(t)$ is distributed according to a Poisson distribution with rate λ, and

$$M(t) = V\{N(t)\} = \frac{tF(t)}{\int_0^t \overline{F}(u) \, du} = \frac{t\overline{F}(t)}{\int_t^\infty \overline{F}(u) \, du} = \lambda t. \qquad (1.25)$$

A Poisson process has stationary independent increments. Eliminating the stationarity, we can generalize a Poisson process with a parameter that is a function of time t as follows:

$$F^{(n)}(t) = \sum_{j=n}^{\infty} \frac{[H(t)]^j}{j!} e^{-H(t)} \qquad (n = 0, 1, 2, \cdots), \qquad (1.26)$$

$$\Pr\{N(t+u) - N(u) = n\} = \frac{[H(t+u) - H(u)]^n}{n!} e^{-[H(t+u)-H(u)]}, \qquad (1.27)$$

$$M(t+u) - M(u) = V\{N(t+u) - N(u)\} = H(t+u) - H(u) \qquad (1.28)$$

for all $u \geq 0$. The counting process $\{N(t), t \geq 0\}$ is called a *nonhomogeneous Poisson process* with a *mean value function* $H(t)$ and $h(t) \equiv dH(t)/dt$ is called an *intensity function*. In addition, from [1, p. 97, 66],

$$E\{X_n\} = \int_0^\infty \frac{[H(t)]^{n-1}}{(n-1)!} e^{-H(t)} dt \qquad (n = 1, 2, \cdots), \qquad (1.29)$$

and if $h(t)$ is increasing, then $E\{X_n\}$ is decreasing in n to $1/h(\infty)$.

Next, suppose that $\{W_j\}$ are independent and identically distributed random variables associated with X_j, and W_j has an identical distribution $G(x)$ with finite mean $E\{W\}$ and is independent of X_i ($i \neq j$), where $W_0 \equiv 0$. When $\{N(t); t \geq 0\}$ is a Poisson process, we consider a new random variable at time t defined by

$$Z(t) \equiv \sum_{j=0}^{N(t)} W_j \qquad (N(t) = 0, 1, 2, \cdots). \qquad (1.30)$$

Then, the stochastic process $\{Z(t), t \geq 0\}$ under two processes is called a *compound Poisson process* [60, 63, 67]. In addition, the LS transform of the distribution of W_j is denoted by $G^*(s) \equiv \int_0^\infty e^{-sx} d\Pr\{W_j \leq x\} = \int_0^\infty e^{-sx} dG(x)$ for $s > 0$. Then, because

$$\Pr\{Z(t) \leq x\} = \sum_{n=0}^\infty \Pr\{W_1 + W_2 + \cdots + W_n \leq x | N(t) = n\} \Pr\{N(t) = n\}$$

$$= \sum_{n=0}^\infty \Pr\{W_1 + W_2 + \cdots + W_n \leq x\} \frac{(\lambda t)^n}{n!} e^{-\lambda t},$$

its LS transform is

$$\int_0^\infty e^{-sx} d\Pr\{Z(t) \leq x\} = \sum_{n=0}^\infty [G^*(s)]^n \frac{(\lambda t)^n}{n!} e^{-\lambda t}$$

$$= \exp\{-\lambda t[1 - G^*(s)]\}. \qquad (1.31)$$

Thus, it easily follows that

$$E\{Z(t)\} = \lambda t E\{W\}, \qquad (1.32)$$

$$V\{Z(t)\} = \lambda t E\{W^2\}. \qquad (1.33)$$

The stochastic process $\{Z(t); t \geq 0\}$ for $\{N(t); t \geq 0\}$ is called a *cumulative process* [11] and some interesting results will be derived in Chapter 2.

(3) Renewal Reward Process

The stochastic process $\{Z(t), t \geq 0\}$, defined in (1.30) when $\{N(t), t \geq 0\}$ is a renewal process, is also called a *renewal reward process* [60]. Using Theorem 1.1,

$$\lim_{t\to\infty} \frac{E\{Z(t)\}}{t} = \frac{E\{W\}}{E\{X\}}, \tag{1.34}$$

where $E\{W\} \equiv E\{W_j\} < \infty$ and $E\{X\} \equiv E\{X_j\} < \infty$ for all $j \geq 1$. This property is applied to the analysis of optimum policies for many maintenance models in reliability theory over an infinite time span [1].

1.2 Shock Processes

Consider a unit subject to damage, wear, and fatigue produced by a series of shocks, jolts, blows, or stresses. When shocks occur in a Poisson process, a renewal process, or in more general stochastic processes, and more simply, at a constant time, the stochastic process $\{Z(t)\}$ defined in (1.30) represents the total cumulative damage at time t.

When shocks occur in a Poisson process, the times between successive shocks are distributed exponentially and has a memoryless property. In other words, shocks are generated randomly and uniformly in time, and the time from any time t to the next shock is independent of time t and has the same exponential distribution as that from time 0. If the unit fails when the total number of shocks has exceeded a specified number n, then the failure time has a gamma distribution given in (1.23).

When shocks occur in a nonhomogeneous Poisson process with an intensity function $h(t)$, the probability that some shock occurs in a small interval $(t, t+dt]$ is given approximately by $h(t)dt$ for any $t \geq 0$. This corresponds to the shock model in which the mean times between shocks decrease with time. For example, consider a two-unit system with failure interaction as described in Section 4.2, in which unit 1 suffers some damage due to the failure of unit 2. If unit 2 undergoes only minimal repair at failures [1, pp. 95–116], then the failure times of unit 2, *i.e.*, shock times of unit 1, are generated according to a nonhomogeneous Poisson process.

Finally, shocks occur in a renewal process, *i.e.*, the sequence of times $\{X_j\}$ between shocks is independent and identically distributed with a general distribution $F(t)$. However, the time $\gamma(t)$ from time t to the next shock, that is called the *excess time* in a stochastic process or *residual lifetime* in reliability theory at time t [61,65], depends on t, and is given by a renewal-type equation

$$\Pr\{\gamma(t) \leq x\} = F(t+x) - \int_0^t [1 - F(t+x-u)]\,dM(u), \tag{1.35}$$

$$\lim_{t\to\infty} \Pr\{\gamma(t) \leq x\} = \frac{1}{E\{X\}} \int_0^x [1 - F(u)]\,du, \tag{1.36}$$

where $E\{X\} \equiv E\{X_j\}$ and $M(t)$ is given in (1.2). This corresponds to the shock model in which a shock will be generated by depending only on the lapse time from the previous shock, regardless of the lapse time of the previous shock.

Furthermore, shocks have been assumed to occur in more generalized stochastic processes such as the birth process [68,69], the Lévy process [70,71], and the general counting process [72,73]. Such studies have given many interesting results theoretically in reliability theory. However, these would not be useful practically for actual reliability models because the contents are too mathematical.

Example 1.1.   Suppose that a unit suffers some damage due to each shock with probability p $(0 < p \leq 1)$ and no damage with probability $q \equiv 1 - p$. We can interpret another example that is the damage of a target hit by a weapon. The probability of hitting a target when a weapon fires at a passive target is p and the probability of missing a target is q. This is called a *stochastic duel* [74,75] and will be dealt with Section 10 as one of related cumulative damage models.

When shocks occur in a renewal process, the distribution of time where the unit suffers some damage for the first time until time t is

$$F_1(t) \equiv [1 + qF(t) + qF(t) * qF(t) + \cdots] * pF(t).$$

Taking the LS transforms on both sides yields

$$F_1^*(s) = \frac{pF^*(s)}{1 - qF^*(s)},$$

and hence, the mean time to the first damage due to some shock is

$$\int_0^\infty t \, dF_1(t) = \frac{E\{X\}}{p}.$$

Thus, by replacing $F(t)$ with $F_1(t)$ in **(1)**, we can get the results in the case where shocks are imperfect. In particular, when $F(t) = 1 - e^{-\lambda t}$, $F_1(t) = 1 - e^{-p\lambda t}$, $\int_0^\infty t dF_1(t) = 1/(p\lambda)$, and

$$\Pr\{N(t) = n\} = \frac{(p\lambda t)^n}{n!} e^{-p\lambda t} \qquad (n = 0, 1, 2, \cdots).$$

Similarly, when shocks occur in a nonhomogeneous Poisson process with a mean value function $H(t)$,

$$\Pr\{N(t) = n\} = \frac{[pH(t)]^n}{n!} e^{-pH(t)} \qquad (n = 0, 1, 2, \cdots),$$

and $E\{N(t)\} = V\{N(t)\} = pH(t)$. ∎

Example 1.2.   Consider a parallel redundant system with n identical units, each of which fails at shocks with probability p $(0 < p \leq 1)$, where $q \equiv 1 - p$, and shocks occur in a renewal process with mean interval μ_1. Let W_j be the total number of units that fail at the jth $(j = 1, 2, \cdots)$ shock. Then, because the probability that one unit fails until the jth shock is

$$\sum_{i=1}^{j} pq^{i-1} = 1 - q^j,$$

the mean time to system failure is [76]

$$\sum_{j=1}^{\infty} j\mu_1 \Pr\left\{W_1 + W_2 + \cdots + W_{j-1} \le n - 1 \text{ and } W_1 + W_2 + \cdots + W_j = n\right\}$$

$$= \sum_{j=1}^{\infty} j\mu_1[(1 - q^j)^n - (1 - q^{j-1})^n]$$

$$= \mu_1 \sum_{j=0}^{\infty} [1 - (1 - q^j)^n]$$

$$= \mu_1 \sum_{i=1}^{n} \binom{n}{i}(-1)^{i+1} \frac{1}{1 - q^i},$$

that is strictly increasing in q from μ_1 to ∞. The replacement problem of this model will be taken up in Section 4.1.1. ∎

More general redundant systems with common-cause failures in which one or more units fail simultaneously at shocks were analyzed [77–80].

2

Damage Models

Consider a standard cumulative damage model [11] for an operating unit: A unit is subjected to shocks and suffers some damage due to shocks. Let random variables X_j $(j = 1, 2, \dots)$ denote a sequence of interarrival times between successive shocks, and random variables W_j $(j = 1, 2, \dots)$ denote the damage produced by the jth shock, where $W_0 \equiv 0$. It is assumed that the sequence of $\{W_j\}$ is nonnegative, independently, and identically distributed, and furthermore, W_j is independent of X_i $(i \neq j)$. This is called a *jump process* [81] or *doubly stochastic process* [82].

Let $N(t)$ denote the random variable that is the total number of shocks up to time t $(t \geq 0)$. Then, define a random variable

$$Z(t) \equiv \sum_{j=0}^{N(t)} W_j \qquad (N(t) = 0, 1, 2, \dots), \qquad (2.1)$$

where $Z(t)$ represents the total damage at time t. It is assumed that the unit fails when the total damage has exceeded a prespecified level K $(0 < K < \infty)$ for the first time (see Figure 2.1). Usually, a failure level K is statistically estimated and is already known. Of interest is a random variable $Y \equiv \min\{t; Z(t) > K\}$, *i.e.*, $\Pr\{Y \leq t\}$ represents the distribution of the failure time of the unit.

In this chapter, we consider two damage models: (1) the cumulative damage model where the total damage is additive, and (2) the independent damage model where the total damage is not additive, *i.e.*, it is independent of the previous damage level. For each model, we are interested in the following reliability quantities:

(i) $\Pr\{Z(t) \leq x\}$; the distribution of the total damage at time t.
(ii) $E\{Z(t)\}$; the total expected damage at time t.
(iii) $\Pr\{Y \leq t\}$; the first-passage time distribution to failure.
(iv) $E\{Y\}$; the mean time to failure (MTTF).

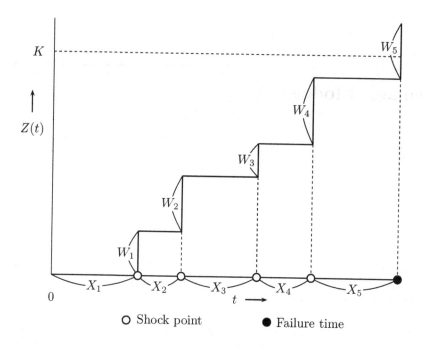

K

$\uparrow$

$Z(t)$

W_5

W_4

W_3

W_2

W_1

X_1 X_2 X_3 X_4 X_5

0

$t \longrightarrow$

O Shock point   ● Failure time

Fig. 2.1. Process for a standard cumulative damage model

(v) Failure rate or hazard rate $r(t)$; $r(t)\mathrm{d}t = \Pr\{t < Y \le t + \mathrm{d}t | Y > t\}$ is the probability that the unit surviving at time t will fail in $(t, t + \mathrm{d}t]$.

(vi) Probability function p_j; p_j is the probability that the unit fails at the jth shock.

Some reliability quantities have already been obtained [11, 33, 40]. This chapter summarizes only the known results that can be applied to maintenance policies discussed in later chapters and be useful in practical fields. A continuous wear process in which the total damage increases with time t is briefly introduced. Finally, five modified damage models are proposed. Several examples are presented. Some examples might appear to be theoretical and contrived, however, these would be useful for understanding the results easily.

2.1 Cumulative Damage Model

Consider a standard cumulative damage model: Successive shocks occur at time intervals X_j $(j = 1, 2, \dots)$ and each shock causes some damage to a unit in the amount W_j. The total damage due to shocks is additive.

It is assumed that $1/\lambda \equiv E\{X_j\} < \infty$, $1/\mu \equiv E\{W_j\} < \infty$, and $F(t) \equiv \Pr\{X_j \le t\}$, $G(x) \equiv \Pr\{W_j \le x\}$ for $t, x \ge 0$. Then, from (1.1) in Chapter 1, the probability that shocks occur exactly j times in $[0, t]$ is [11]

$$\Pr\{N(t) = j\} = F^{(j)}(t) - F^{(j+1)}(t) \qquad (j = 0, 1, 2, \dots).$$

Thus,

$$\Pr\left\{\sum_{i=0}^{N(t)} W_i \leq x, N(t) = j\right\} = \Pr\left\{\sum_{i=0}^{N(t)} W_i \leq x \middle| N(t) = j\right\} \Pr\{N(t) = j\}$$

$$= G^{(j)}(x)[F^{(j)}(t) - F^{(j+1)}(t)] \qquad (j = 0, 1, 2, \dots), \qquad (2.2)$$

where $\varphi^{(j)}(t)$ denotes the j-fold Stieltjes convolution of any function $\varphi(t)$ with itself, and $\varphi^{(0)}(t) \equiv 1$ for $t \geq 0$.

Therefore, the distribution of $Z(t)$ defined in (2.1) is

$$\Pr\{Z(t) \leq x\} = \Pr\left\{\sum_{i=0}^{N(t)} W_i \leq x\right\}$$

$$= \sum_{j=0}^{\infty} \Pr\left\{\sum_{i=0}^{N(t)} W_i \leq x \middle| N(t) = j\right\} \Pr\{N(t) = j\}$$

$$= \sum_{j=0}^{\infty} G^{(j)}(x)[F^{(j)}(t) - F^{(j+1)}(t)], \qquad (2.3)$$

and the survival probability is

$$\Pr\{Z(t) > x\} = \sum_{j=0}^{\infty} [G^{(j)}(x) - G^{(j+1)}(x)]F^{(j+1)}(t). \qquad (2.4)$$

The total expected damage at time t is

$$E\{Z(t)\} = \int_0^{\infty} x \, d\Pr\{Z(t) \leq x\}$$

$$= \frac{1}{\mu} \sum_{j=1}^{\infty} F^{(j)}(t) = \frac{M_F(t)}{\mu}, \qquad (2.5)$$

where $M_F(t) \equiv \sum_{j=1}^{\infty} F^{(j)}(t)$ is called a renewal function of distribution $F(t)$ and represents the expected number of shocks in $[0, t]$. It can be intuitively known that $E\{Z(t)\}$ is given by the product of the average amount of damage suffered from shocks and the expected number of shocks in time t. This is useful for estimating the total expected damage at time t.

Furthermore, from Theorem 1.2, for the distribution F with finite rth moment μ_r and variance σ^2,

$$M(t) \equiv E\{N(t)\} = \frac{t}{\mu_1} + \left(\frac{\sigma^2}{2\mu_1^2} - \frac{1}{2}\right) + o(1),$$

$$V\{N(t)\} = \frac{\sigma^2 t}{\mu_1^3} + \left(\frac{5\sigma^4}{4\mu_1^4} + \frac{2\sigma^2}{\mu_1^2} + \frac{3}{4} - \frac{2\mu_3}{3\mu_1^3}\right) + o(1).$$

Thus, when F (G) has finite mean $1/\lambda$ $(1/\mu)$ and variance σ_F^2 (σ_G^2), approximately, for large t,

$$
E\{Z(t)\} = E\left\{E\left\{\sum_{j=1}^{N(t)} W_j \Big| N(t)\right\}\right\} = E\{N(t)\}E\{W_j\}
$$

$$
\approx \frac{1}{\mu}\left(\lambda t + \frac{\lambda^2\sigma_F^2 - 1}{2}\right),
\tag{2.6}
$$

$$
V\{Z(t)\} = E\{Z^2(t)\} - [E\{Z(t)\}]^2
$$

$$
= E\left\{\left\{\sum_{j=1}^{N(t)} W_j \sum_{i=1}^{N(t)} W_i \Big| N(t)\right\}\right\} - [E\{Z(t)\}]^2
$$

$$
= V\{N(t)\}[E\{W_j\}]^2 + E\{N(t)\}V\{W_j\}
$$

$$
\approx \frac{1}{\mu}\left[\frac{\lambda t}{\mu}(\lambda^2\sigma_F^2 + \mu^2\sigma_G^2) + \frac{1}{\mu}\left(\frac{5\lambda^4\sigma_F^4}{4} + 2\lambda^2\sigma_F^2 + \frac{3}{4} - \frac{2\lambda^3\mu_3}{3}\right)\right]
$$

$$
+ \frac{\sigma_G^2}{2}(\lambda^2\sigma_F^2 - 1).
\tag{2.7}
$$

Moreover, because

$$
\lim_{t\to\infty} \frac{E\{Z(t)\}}{t} = \frac{\lambda}{\mu}, \qquad \lim_{t\to\infty} \frac{V\{Z(t)\}}{t} = \frac{\lambda}{\mu^2}(\lambda^2\sigma_F^2 + \mu^2\sigma_G^2),
$$

by applying Takács theorem [83] (see Example 2.6 in [1]) to this model,

$$
\lim_{t\to\infty} \Pr\left\{\frac{Z(t) - \lambda t/\mu}{\sqrt{\lambda^3 t(\sigma_F^2/\mu^2 + \sigma_G^2/\lambda^2)}} \le x\right\} = \frac{1}{\sqrt{2\pi}}\int_{-\infty}^{x} e^{-u^2/2}\,du.
\tag{2.8}
$$

This was proved in [29] and generalized in [84–86].

Example 2.1.   We wish to estimate the total damage when the probability that it is more than z in $t = 30$ days of operation is given by 0.90. The distributions of shock times and the amount of damage are unknown, but from sample data, the following estimations of means and variances are made:

$$
\begin{aligned}
1/\lambda &= 2\,\text{days}, & \sigma_F^2 &= 5\,(\text{days})^2, \\
1/\mu &= 1, & \sigma_G^2 &= 0.5.
\end{aligned}
$$

In this case, from (2.6), $E\{Z(30)\} \approx 15.125$. Then, from (2.8), when $t = 30$,

$$
\frac{Z(t) - \lambda t/\mu}{\sqrt{\lambda^3 t(\sigma_F^2/\mu^2 + \sigma_G^2/\lambda^2)}} = \frac{Z(30) - 15}{5.12}
$$

is approximately normally distributed with mean 0 and variance 1. Hence,

$$\Pr\{Z(t) > z\} = \Pr\left\{\frac{Z(30) - 15}{5.12} > \frac{z - 15}{5.12}\right\}$$

$$\approx \frac{1}{\sqrt{2\pi}} \int_{(z-15)/5.12}^{\infty} e^{-u^2/2} \, du = 0.90.$$

Because $u_0 = -1.28$ such that $(1/\sqrt{2\pi}) \int_{u_0}^{\infty} e^{-u^2/2} \, du = 0.90$, $z = 15 - 5.12 \times 1.28 \approx 8.45$. Thus, the total damage is more than 8.45 in 30 days with probability 0.90.

Next, when a failure level is known as $K = 10$,

$$\Pr\{Z(t) > 10\} = \Pr\left\{\frac{Z(30) - 10}{5.12} > \frac{10 - 15}{5.12}\right\}$$

$$\approx \frac{1}{\sqrt{2\pi}} \int_{-0.98}^{\infty} e^{-u^2/2} \, du \approx 0.84.$$

Thus, the probability that the unit with a failure level $K = 10$ fails in 30 days is about 0.84. $\blacksquare$

The first-passage time distribution to failure when the failure level is constant K, because the events of $\{Y \leq t\}$ and $\{Z(t) > K\}$ are equivalent, is, from (2.4),

$$\Phi(t) \equiv \Pr\{Y \leq t\} = \Pr\{Z(t) > K\}$$

$$= \sum_{j=0}^{\infty} [G^{(j)}(K) - G^{(j+1)}(K)] F^{(j+1)}(t), \tag{2.9}$$

and its Laplace–Stieltjes (LS) transform is

$$\Phi^*(s) \equiv \int_0^{\infty} e^{-st} \, d\Phi(t) = \sum_{j=0}^{\infty} [G^{(j)}(K) - G^{(j+1)}(K)][F^*(s)]^{j+1}, \tag{2.10}$$

where $\varphi^*(s)$ denotes the LS transform of any function $\varphi(t)$, i.e., $\varphi^*(s) \equiv \int_0^{\infty} e^{-st} d\varphi(t)$ for $s > 0$. Thus, the mean time to failure is

$$E\{Y\} = \int_0^{\infty} t \, d\Pr\{Y \leq t\} = -\left.\frac{d\Phi^*(s)}{ds}\right|_{s=0}$$

$$= \frac{1}{\lambda} \sum_{j=0}^{\infty} G^{(j)}(K) = \frac{1}{\lambda}[1 + M_G(K)], \tag{2.11}$$

where $M_G(K) \equiv \sum_{j=1}^{\infty} G^{(j)}(K)$ represents the expected number of shocks before the total damage exceeds a failure level K.

Similarly, when G has finite mean $1/\mu$ and variance σ_G^2, approximately,

$$E\{Y\} \approx \frac{1}{\lambda}\left(\mu K + \frac{\mu^2 \sigma_G^2 + 1}{2}\right). \tag{2.12}$$

In addition, when the distribution G has an IFR property, it has been shown that $\mu x - 1 < M_G(x) \le \mu x$ from (1.20). Thus,

$$\frac{\mu K}{\lambda} < E\{Y\} \le \frac{\mu K + 1}{\lambda}. \tag{2.13}$$

In Example 2.1, $E\{Y\}$ is approximately 21.5 days and $20 < E\{Y\} \le 22$.
Finally, the failure rate is

$$
\begin{aligned}
r(t)\,dt &= \frac{\Pr\{t < Y \le t + dt\}}{\Pr\{Y > t\}} \\
&= \frac{\sum_{j=0}^{\infty}[G^{(j)}(K) - G^{(j+1)}(K)]f^{(j+1)}(t)\,dt}{\sum_{j=0}^{\infty} G^{(j)}(K)[F^{(j)}(t) - F^{(j+1)}(t)]},
\end{aligned} \tag{2.14}
$$

where $f(t)$ is a density function of $F(t)$. Furthermore, because the probability that the unit fails at the $(j + 1)$th shock is $p_{j+1} \equiv G^{(j)}(K) - G^{(j+1)}(K)$ $(j = 0, 1, 2, \dots)$, its survival distribution is

$$\overline{P}_j \equiv \sum_{i=j}^{\infty} p_{i+1} = G^{(j)}(K) \qquad (j = 0, 1, 2, \dots),$$

where $\overline{P}_0 \equiv 1$, i.e., $\overline{P}_j$ represents the probability of surviving the first j shocks. Thus, the expected number of shocks until failure, including the shock at which the unit has failed, is

$$\sum_{j=1}^{\infty} j p_j = \sum_{j=0}^{\infty} G^{(j)}(K) = 1 + M_G(K).$$

$E\{Y\}$ in (2.11) is given by the product of the mean time between successive shocks and the expected number of shocks until the total damage has exceeded K. It is also approximately

$$\sum_{j=1}^{\infty} j p_j \approx \mu K + \frac{\mu^2 \sigma_G^2 + 1}{2}.$$

The discrete failure rate for a probability function $\{p_j\}_{j=1}^{\infty}$ is

$$r_{j+1} \equiv \frac{p_{j+1}}{\overline{P}_j} = \frac{G^{(j)}(K) - G^{(j+1)}(K)}{G^{(j)}(K)} \qquad (j = 0, 1, 2, \dots), \tag{2.15}$$

i.e., r_{j+1} represents the probability that the unit surviving at the jth shock will fail at the $(j + 1)$th shock and is less than or equal to 1.

Next, suppose that shocks occur in a nonhomogeneous Poisson process with an intensity function $h(t)$ and a mean value function $H(t)$, i.e., $H(t) \equiv \int_0^t h(u)\,du$ in (2) of Section 1.1. Then, from (1.1) and (1.26),

$$\Pr\{N(t) = j\} = \frac{[H(t)]^j}{j!}e^{-H(t)} \qquad (j = 0, 1, 2, \dots). \qquad (2.16)$$

Thus, by replacing $F^{(j)}(t)$ with $\sum_{i=j}^{\infty}\{[H(t)]^i/i!\}e^{-H(t)}$ formally, we can rewrite all reliability quantities. For example,

$$\Pr\{Z(t) \le x\} = \sum_{j=0}^{\infty} G^{(j)}(x)\frac{[H(t)]^j}{j!}e^{-H(t)}, \qquad (2.17)$$

$$E\{Z(t)\} = \frac{H(t)}{\mu}, \qquad (2.18)$$

$$E\{Y\} = \sum_{j=0}^{\infty} G^{(j)}(K) \int_0^{\infty} \frac{[H(t)]^j}{j!}e^{-H(t)}\,dt. \qquad (2.19)$$

If shocks occur at a constant time t_0 $(0 < t_0 < \infty)$, i.e., $F(t)$ is the degenerate distribution placing unit mass at time t_0, and $F(t) \equiv 0$ for $t < t_0$, and 1 for $t \ge t_0$, then

$$\Pr\{Y \le t\} = 1 - G^{([t/t_0])}(K),$$

$$E\{Y\} = \int_0^{\infty} G^{([t/t_0])}(K)\,dt,$$

where $[t/t_0]$ denotes the greatest integer less than or equal to t/t_0.

Finally, when $G(x) \equiv 0$ for $x < 1$ and 1 for $x \ge 1$, and $K = n$,

$$\Pr\{Y \le t\} = F^{(n+1)}(t), \qquad E\{Y\} = \frac{n+1}{\lambda},$$

that is, the unit fails certainly at the $(n + 1)$th shock.

2.2 Independent Damage Model

Consider the independent damage model for an operating unit where the total damage is not additive, i.e., any shock does no damage unless its amount has not exceeded a failure level K. If the damage due to some shock has exceeded for the first time a failure level K, then the unit fails (see Figure 2.2). The same assumptions as those of the previous model are made except that the total damage is additive. A typical example of this model is the fracture of brittle materials such as glasses [33], and semiconductor parts that have failed by some overcurrent or fault voltage. The generalized model with three types of shocks where shocks with a small level of damage are no damage to the unit, shocks with a large level of damage result in failure, and shocks with an intermediate level result in failure only with some probability, was considered [87].

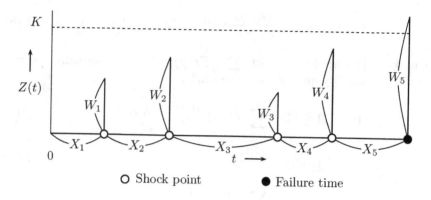

O Shock point   ● Failure time

Fig. 2.2. Process for an independent damage model

In this case, the probability that the unit fails exactly at the $(j + 1)$th shock $(j = 0, 1, 2, \dots)$ is $p_{j+1} = [G(K)]^j - [G(K)]^{j+1}$. Thus, the distribution of time to failure is

$$\Pr\{Y \le t\} = \sum_{j=0}^{\infty} \{[G(K)]^j - [G(K)]^{j+1}\} F^{(j+1)}(t), \qquad (2.20)$$

its LS transform is

$$\int_0^{\infty} e^{-st} \, d\Pr\{Y \le t\} = \frac{[1 - G(K)]F^*(s)}{1 - G(K)F^*(s)}, \qquad (2.21)$$

and the mean time to failure is

$$E\{Y\} = \frac{1}{\lambda[1 - G(K)]}. \qquad (2.22)$$

Furthermore, the failure rates are

$$r(t) = \frac{\sum_{j=0}^{\infty} \{[G(K)]^j - [G(K)]^{j+1}\} f^{(j+1)}(t)}{\sum_{j=0}^{\infty} [G(K)]^j [F^{(j)}(t) - F^{(j+1)}(t)]}, \qquad (2.23)$$

$$r_{j+1} = p_1 = 1 - G(K) \qquad (j = 0, 1, 2, \dots), \qquad (2.24)$$

that is constant for any j.

If shocks occur in a nonhomogeneous Poisson process with a mean value function $H(t)$, then,

$$\Pr\{Y \le t\} = \sum_{j=0}^{\infty} \{1 - [G(K)]^j\} \frac{[H(t)]^j}{j!} e^{-H(t)} = 1 - e^{-[1-G(K)]H(t)}, \quad (2.25)$$

and its mean time is

$$E\{Y\} = \int_0^\infty e^{-[1-G(K)]H(t)}\, dt. \tag{2.26}$$

The failure rate is

$$r(t) = [1 - G(K)]h(t), \tag{2.27}$$

that has the same property as that of an intensity function $h(t)$.

If shocks occur at a constant time t_0,

$$\Pr\{Y \le t\} = 1 - [G(K)]^{[t/t_0]},$$

$$E\{Y\} = \int_0^\infty [G(K)]^{[t/t_0]}\, dt.$$

Example 2.2. Suppose that $F(t) = 1 - e^{-\lambda t}$ and $G(x) = 1 - e^{-\mu x}$, *i.e.*, shocks occur in a Poisson process with rate λ and each damage due to shocks is exponential with mean $1/\mu$. In this case, both a nonhomogeneous Poisson and renewal processes form the same Poisson process, *i.e.*,

$$F^{(j)}(t) = \sum_{i=j}^\infty \frac{[H(t)]^i}{i!} e^{-H(t)} = \sum_{i=j}^\infty \frac{(\lambda t)^i}{i!} e^{-\lambda t} \qquad (j = 0, 1, 2, \dots).$$

In the cumulative damage model of Section 2.1, from (1.31),

$$\int_0^\infty e^{-sx}\, d\Pr\{Z(t) \le x\} = e^{-\lambda[s/(s+\mu)t]}.$$

By inversion [65, p. 80],

$$\Pr\{Z(t) \le x\} = e^{-\lambda t}\left[1 + \sqrt{\lambda\mu t}\int_0^x e^{-\mu u} u^{-1/2} I_1\left(2\sqrt{\lambda\mu t u}\right) du\right],$$

where $I_i(x)$ is the Bessel function of order i for the imaginary argument defined by

$$I_i(x) \equiv \sum_{j=0}^\infty \left(\frac{x}{2}\right)^{2j+i} \frac{1}{j!(j+i)!}.$$

Thus, from (2.9), the distribution of time to failure is

$$\Pr\{Y \le t\} = 1 - e^{-\lambda t}\left[1 + \sqrt{\lambda\mu t}\int_0^K e^{-\mu u} u^{-1/2} I_1\left(2\sqrt{\lambda\mu t u}\right) du\right].$$

Furthermore, from (2.5), (2.11), or (2.18), (2.19), and (2.7),

$$E\{Z(t)\} = \frac{\lambda t}{\mu}, \qquad V\{Z(t)\} = \frac{2\lambda t}{\mu^2},$$

$$E\{Y\} = \frac{1}{\lambda}\sum_{j=1}^\infty j p_j = \frac{\mu K + 1}{\lambda},$$

where note that $E\{Z(t)\}$ increases linearly with time t. Thus, we have the interesting result

$$\frac{E\{Z(t)\}}{K + 1/\mu} = \frac{t}{E\{Y\}},$$

that represents that the ratio of the total expected damage at time t to a failure level plus one mean amount of damage is equal to that of the time t to the mean time to failure. If the mean time between shock times and their mean damage due to shocks are roughly estimated, the mean damage level and the mean time to failure are also estimated easily from these relations.

The failure rates are, from (2.14) and (2.15), respectively,

$$r(t) = \frac{\lambda e^{-\lambda t - \mu K} I_0 \left(2\sqrt{\lambda \mu t K}\right)}{1 + \sqrt{\lambda \mu t} \int_0^K e^{-\mu u} u^{-1/2} I_1 \left(2\sqrt{\lambda \mu t u}\right) du},$$

$$r_{j+1} = \frac{(\mu K)^j / j!}{\sum_{i=j}^{\infty} [(\mu K)^i / i!]} \qquad (j = 0, 1, 2, \dots),$$

that is strictly increasing in j from $e^{-\mu K}$ to 1, because

$$r_{j+1} - r_j = \frac{(\mu K)^j / j!}{\sum_{i=j}^{\infty} [(\mu K)^i / i!]} - \frac{(\mu K)^{j-1} / (j-1)!}{\sum_{i=j-1}^{\infty} [(\mu K)^i / i!]}$$

$$= \frac{\sum_{i=j}^{\infty} [(\mu K)^{i+j-1} / (i! j!)] (i - j)}{\sum_{i=j}^{\infty} [(\mu K)^i / i!] \sum_{i=j-1}^{\infty} [(\mu K)^i / i!]} > 0.$$

In the independent damage model of Section 2.2, from (2.20) or (2.25),

$$\Pr\{Y \le t\} = 1 - \exp(-\lambda t e^{-\mu K}),$$

and from (2.22) or (2.26),

$$E\{Y\} = \frac{1}{r(t)} = \frac{1}{\lambda} e^{\mu K},$$

that is, the first-passage time Y to failure has an exponential distribution with mean $e^{\mu K}/\lambda$ and the failure rate is constant. ∎

2.3 Failure Rate

Investigate the reliability properties of the survival distribution $\overline{\Phi}(t) \equiv 1 - \Phi(t) = \Pr\{Y > t\}$ that the unit does not fail in $[0, t]$. Let $\overline{P}_j$ denote the probability of surviving the first j shocks $(j = 0, 1, 2, \dots)$, where $P_0 \equiv 0$, and $F_j(t)$ be the probability that j shocks occur in time t, where $F_0(t) \equiv 1$. Then, the survival distribution is written in the following general form:

$$\overline{\Phi}(t) = \sum_{j=0}^{\infty} \overline{P}_j \Pr\{N(t) = j\} = \sum_{j=0}^{\infty} \overline{P}_j [F_j(t) - F_{j+1}(t)]. \qquad (2.28)$$

In particular, when shocks occur in a Poisson process with rate $\lambda > 0$, *i.e.*, $F(t) = 1 - e^{-\lambda t}$ in Section 2.1,

$$\overline{\Phi}(t) = \sum_{j=0}^{\infty} \overline{P}_j \frac{(\lambda t)^j}{j!} e^{-\lambda t}. \qquad (2.29)$$

The probabilistic properties of $\overline{\Phi}(t)$ were extensively investigated [34, 88]. We refer briefly only to these results that will be needed in the following chapters: The failure rate is, from (2.14),

$$r(t) = \lambda \left\{ 1 - \frac{\sum_{j=0}^{\infty} \overline{P}_{j+1}[(\lambda t)^j / j!]}{\sum_{j=0}^{\infty} \overline{P}_j[(\lambda t)^j / j!]} \right\} \leq \lambda. \qquad (2.30)$$

When $\overline{P}_j = q^j$, *i.e.*, the total damage is not additive in Section 2.2, $\overline{\Phi}(t) = e^{-\lambda(1-q)t}$ and $r(t) = \lambda(1-q)$ is constant.

Any distribution $F(t)$ is said to have the property of IFR (increasing failure rate) or IHR (increasing hazard rate) if and only if $[F(t+x) - F(t)]/\overline{F}(t)$ is increasing in t for $x > 0$ and $F(t) < 1$ [65], where $\overline{F}(t) \equiv 1 - F(t)$. Furthermore, it has been proved that $F(t)$ is IFR if and only if $r(t) \equiv f(t)/\overline{F}(t)$ is increasing in t. In this model, the following properties (i) and (ii) were proved [33]:

(i) The failure rate $r(t)$ in (2.30) is increasing if $(\overline{P}_j - \overline{P}_{j+1})/\overline{P}_j$ is increasing in j.

In addition, when the total damage is additive and shocks times are exponential, from (2.29),

$$\overline{\Phi}(t) = \sum_{j=0}^{\infty} G^{(j)}(K) \frac{(\lambda t)^j}{j!} e^{-\lambda t}. \qquad (2.31)$$

(ii) The failure rate average $\int_0^t r(u)\mathrm{d}u/t$ is increasing in t because $[G^{(j)}(x)]^{1/j}$ is decreasing in j. Note that if $r(t)$ is increasing, then $\int_0^t r(u)\mathrm{d}u/t$ is also increasing.

In particular, when $\overline{P}_j = G^{(j)}(K) = \sum_{i=j}^{\infty} [(\mu K)^i / i!] e^{-\mu K}$, $\overline{P}_{j+1}/\overline{P}_j$ is strictly decreasing from Example 2.2, so that the failure rate $r(t)$ in (2.30) is strictly increasing from $\lambda e^{-\mu K}$ to λ.

When shocks occur in a nonhomogeneous Poisson process with an intensity function $h(t)$ and a mean value function $H(t)$ [89], from (2.28),

$$\overline{\Phi}(t) = \sum_{j=0}^{\infty} \overline{P}_j \frac{[H(t)]^j}{j!} e^{-H(t)}. \qquad (2.32)$$

(iii) The failure rate $r(t)$ is increasing if $h(t)$ is increasing and $(\overline{P}_j - \overline{P}_{j+1})/\overline{P}_j$ is increasing.

(iv) The failure rate average $\int_0^t r(u)du/t$ is increasing if both $H(t)/t$ and $(\overline{P}_j - \overline{P}_{j+1})/\overline{P}_j$ are increasing.

When the total damage is additive, (2.32) is

$$\overline{\Phi}(t) = \sum_{j=0}^{\infty} G^{(j)}(K) \frac{[H(t)]^j}{j!} e^{-H(t)}. \tag{2.33}$$

Then, properties (iii) and (iv) are rewritten as:

(v) The failure rate $r(t)$ is increasing if $h(t)$ is increasing and r_{j+1} in (2.15) is increasing.

(vi) The failure rate average $\int_0^t r(u)du/t$ is increasing if both $H(t)/t$ and r_{j+1} are increasing.

Such results were compactly summarized [90]. Moreover, when shocks occur in the birth process [68], in the counting process [72], and in the Lévy process [70], similar results were obtained.

After that, damage or shock models of this kind have been generalized and analyzed by many authors [91–107]. A general shock model, where the amount of damage due to shocks is correlated with their intervals, was analyzed [108–114]. Furthermore, bivariate and multivariate distributions derived from cumulative damage models were studied [115–123]. The failure rate was investigated for point, alternating, and diffused stresses [124].

2.4 Continuous Wear Processes

Let Y be the failure time of an operating unit. It is assumed that there exists a nonnegative function $h(t)$ such that

$$\Pr\{t < Y \leq t + \Delta t\} = h(t)\Delta t + o(\Delta t) \tag{2.34}$$

for $\Delta t > 0$ and $t \geq 0$. Then, the probability of the unit surviving at time t is

$$R(t) = \Pr\{Y > t\} = \exp\left[-\int_0^t h(u)\,du\right] = e^{-H(t)}, \tag{2.35}$$

that represents the reliability of the unit at time t and is given in (1.1) of [1]. In this case, the function $h(t)$ is called an *instantaneous wear* and $H(t) \equiv \int_0^t h(u)\,du$ is called an *accumulated wear* at time t [37]. In particular, when $H(t) = at/K$ for $a > 0$, $R(t) = e^{-at/K}$ and $E\{Y\} = K/a$. Furthermore, when $H(t) = \lambda t^m$ $(m > 0)$, $R(t)$ becomes a Weibull distribution and $R(t) = \exp(-\lambda t^m)$.

On the other hand, assume that $h(t)$ is the realization of the stochastic process $\{W(t), t \geq 0\}$ with independent increments [35]. Then,

$$R(t) = E\left\{\exp\left[-\int_0^t W(u)\,du\right]\right\}.$$

(2.36)

If $Z(t)$ is simply the accumulated wear in a stochastic process with independent increments, then [34]

$$R(t) = E\{e^{-Z(t)}\}.$$

(2.37)

The reliability function $R(t)$ was given by a gamma distribution [125] and some reliability functions were derived in more general assumptions [126].

The accumulated wear function $Z(t)$ usually increases with time t from 0, and the unit fails when $Z(t)$ has exceeded a failure level K. Next, suppose that $Z(t) = A_t t + B_t$ for $A_t \geq 0$. Then, the reliability at time t is

$$R(t) = \Pr\{Z(t) \leq K\} = \Pr\{A_t t + B_t \leq K\}.$$

(2.38)

(1) When $A_t \equiv a$ (constant), $K \equiv k$ (constant), and B_t is distributed normally with mean 0 and variance $\sigma^2 t$,

$$R(t) = \Pr\{B_t \leq k - at\} = \Phi\left(\frac{k - at}{\sigma\sqrt{t}}\right),$$

(2.39)

where $\Phi(x)$ is the standard normal distribution with mean 0 and variance 1, i.e., $\Phi(x) = (1/\sqrt{2\pi})\int_{-\infty}^x e^{-u^2/2}du$.

(2) When $B_t \equiv 0$, $K \equiv k$, and A_t is distributed normally with mean a and variance σ^2/t,

$$R(t) = \Pr\{A_t \leq k/t\} = \Phi\left(\frac{k - at}{\sigma\sqrt{t}}\right),$$

(2.40)

that becomes equal to (2.39).

(3) When $A_t \equiv a$, $B_t \equiv 0$, and K is distributed normally with mean k and variance σ^2,

$$R(t) = \Pr\{at \leq K\} = \Phi\left(\frac{k - at}{\sigma}\right).$$

(2.41)

When K is distributed normally with mean k and variance $\sigma^2 t$, $R(t)$ is equal to (2.39) and (2.40).

Replacing $\alpha \equiv \sigma/\sqrt{ak}$ and $\beta \equiv k/a$ in (2.39) or (2.40),

$$R(t) = \Phi\left[\frac{1}{\alpha}\left(\sqrt{\frac{\beta}{t}} - \sqrt{\frac{t}{\beta}}\right)\right],$$

(2.42)

that is called the Birnbaum–Saunders distribution [36, 127]. This is widely applied to fatigue failure for material strength subject to stresses [128–130].

When $Z(t) = \mu t + \sigma B_t$ with positive drift μ and variance σ^2 where B_t is a standard Brownian motion, $Z(t)$ forms the Wiener process or Brownian motion process [62]. However, this has not been applied to actual damage models. When $Z(t) = A_t t + B_t$, if A_t, B_t and K are deterministic, $i.e.$, $A_t \equiv a$, $B_t \equiv b$, and $K \equiv k$, then the unit fails at time $t = (k - b)/a$. By fitting appropriate distributions to A_t, B_t, and K and estimating their parameters for practical systems, the function $Z(t)$ can be used as a continuous wear function in cumulative damage models. When $Z(t) = at$ and K is a random variable, the optimum policy where the unit is replaced at a planned time will be discussed in Section 5.2.

2.5 Modified Damage Models

Let us consider the following five damage models mainly based on our own work: (1) damage model with imperfect shock where some shock may produce no damage to a unit [40], (2) a failure level is a random variable with a general distribution $L(x)$ [131], (3) the total damage decreases exponentially with time [132], (4) the damage model of a system with n different units [133], and (5) the total damage increases with time [14, 134, 135]. Such damage models would be realistic in reliability models and be useful in practice. We derive the reliability quantities of each model and show simple examples when shock times are exponential.

(1) Imperfect Shock

It has been assumed that the damage due to a shock occurs and its amount is distributed with $G(x)$. However, it may be considered that some shocks do not produce any damage to a unit.

Suppose that the damage due to shocks occurs with probability p ($0 < p \le 1$) and does not occur with probability $q \equiv 1 - p$. Other notations are the same as those of Sections 2.1 and 2.2. Then, substituting $F_1(t)$ in Example 1.1 in $F(t)$ in (2.3), (2.5), (2.9), (2.11), and (2.14), $\Pr\{Z(t) \le x\}$, $E\{Z(t)\}$, $\Pr\{Y \le t\}$, $E\{Y\}$, and $r(t)$ are given. In particular, from (2.10) and (2.11), respectively,

$$\int_0^\infty e^{-st}\, d\Pr\{Y \le t\} = \sum_{j=0}^\infty [G^{(j)}(K) - G^{(j+1)}(K)] \left[\frac{pF^*(s)}{1 - qF^*(s)}\right]^{j+1}, \quad (2.43)$$

$$E\{Y\} = \frac{1}{p\lambda} \sum_{j=0}^\infty G^{(j)}(K) = \frac{1}{p\lambda}[1 + M_G(K)]. \quad (2.44)$$

The corresponding results for the independent damage model are, from (2.21) and (2.22), respectively,

$$\int_0^\infty e^{-st}\,\mathrm{dPr}\{Y \le t\} = \frac{p[1 - G(K)]F^*(s)}{1 - [q + pG(K)]F^*(s)}, \tag{2.45}$$

$$E\{Y\} = \frac{1}{p\lambda[1 - G(K)]}. \tag{2.46}$$

(2) Random Failure Level and Time-Dependent Failure Level

Most units have individual variations in their ability to withstand shocks and are operating in a different environment. In such cases, a failure level K is not constant and would be random. Consider the case where a failure level K is a random variable with a general distribution $L(x)$ such that $L(0) = 0$ [33]. Then, for the cumulative damage model, the distribution of time to failure is

$$\mathrm{Pr}\{Y \le t\} = \sum_{j=0}^\infty F^{(j+1)}(t) \int_0^\infty [G^{(j)}(x) - G^{(j+1)}(x)]\,\mathrm{d}L(x), \tag{2.47}$$

and its mean time is

$$E\{Y\} = \frac{1}{\lambda} \sum_{j=0}^\infty \int_0^\infty G^{(j)}(x)\,\mathrm{d}L(x). \tag{2.48}$$

The failure rates are

$$r(t) = \frac{\sum_{j=0}^\infty f^{(j+1)}(t) \int_0^\infty [G^{(j)}(x) - G^{(j+1)}(x)]\,\mathrm{d}L(x)}{\sum_{j=0}^\infty [F^{(j)}(t) - F^{(j+1)}(t)] \int_0^\infty G^{(j)}(x)\,\mathrm{d}L(x)}, \tag{2.49}$$

$$r_{j+1} = \frac{\int_0^\infty [G^{(j)}(x) - G^{(j+1)}(x)]\,\mathrm{d}L(x)}{\int_0^\infty G^{(j)}(x)\,\mathrm{d}L(x)}. \tag{2.50}$$

For the independent damage model,

$$\mathrm{Pr}\{Y \le t\} = \sum_{j=0}^\infty F^{(j+1)}(t) \int_0^\infty \{[G(x)]^j - [G(x)]^{j+1}\}\,\mathrm{d}L(x), \tag{2.51}$$

$$E\{Y\} = \frac{1}{\lambda} \sum_{j=0}^\infty \int_0^\infty [G(x)]^j\,\mathrm{d}L(x). \tag{2.52}$$

For the cumulative model with imperfect shock,

$$\int_0^\infty e^{-st}\,\mathrm{dPr}\{Y \le t\} = \sum_{j=0}^\infty \left[\frac{pF^*(s)}{1 - qF^*(s)}\right]^{j+1} \int_0^\infty [G^{(j)}(x) - G^{(j+1)}(x)]\,\mathrm{d}L(x). \tag{2.53}$$

Example 2.3. Suppose that all random variables are exponential, *i.e.*, $F(t) = 1 - e^{-\lambda t}$ and $G(x) = 1 - e^{-\mu x}$. Then, we obtain the explicit formulas for each model.

For imperfect shock, $F_1^*(s) = p\lambda/(s + p\lambda)$, i.e., $F_1(t) = 1 - e^{-p\lambda t}$ by inversion. Thus, substituting λ in $p\lambda$ in Example 2.2, we can obtain the corresponding results.

When a failure level $L(x)$ has also an exponential distribution $(1 - e^{-\theta x})$,

$$\int_0^\infty [G^{(j)}(x) - G^{(j+1)}(x)]\,dL(x) = \frac{\theta\mu^j}{(\mu + \theta)^{j+1}}.$$

Thus, from (2.47),

$$\int_0^\infty e^{-st}\,d\Pr\{Y \le t\} = \sum_{j=0}^\infty \left(\frac{\lambda}{s + \lambda}\right)^{j+1} \frac{\theta\mu^j}{(\mu + \theta)^{j+1}} = \frac{\lambda\theta}{s(\mu + \theta) + \lambda\theta}.$$

By inversion,

$$\Pr\{Y \le t\} = 1 - \exp\left(-\frac{\lambda\theta t}{\mu + \theta}\right),$$

$$E\{Y\} = \frac{1}{r(t)} = \frac{1}{\lambda}\sum_{j=1}^\infty jp_j = \frac{1}{\lambda}\left(\frac{\mu}{\theta} + 1\right),$$

$$r_{j+1} = \frac{\theta}{\mu + \theta} = \frac{r(t)}{\lambda}.$$

It is of great interest that both failure rates are constant, and r_j corresponds to the ratio of (mean damage of one shock)/(mean failure level + mean damage of one shock).

For the independent damage model,

$$\Pr\{Y > t\} = \int_0^\infty \exp(-\lambda t e^{-\mu x})\theta e^{-\theta x}\,dx = \sum_{j=0}^\infty \frac{(-\lambda t)^j}{j!}\int_0^\infty \theta e^{-(\theta + j\mu)x}\,dx$$

$$= \sum_{j=0}^\infty \frac{(-\lambda t)^j}{j!}\frac{\theta}{\theta + j\mu},$$

$$E\{Y\} = \frac{1}{r(t)} = \frac{1}{\lambda}\sum_{j=1}^\infty jp_j$$

$$= \frac{1}{\lambda}\int_0^\infty e^{\mu x}\theta e^{-\theta x}\,dx = \begin{cases} \dfrac{\theta}{\lambda(\theta - \mu)} & (\theta > \mu), \\ \infty & (\theta \le \mu). \end{cases}\quad\blacksquare$$

Finally, suppose that the total damage due to shocks is investigated and is known statistically at the beginning. Then, if the unit with damage z_0 $(0 \le z_0 < K)$ begins to operate at time 0, we can obtain all reliability quantities by replacing K with $K - z_0$ [136].

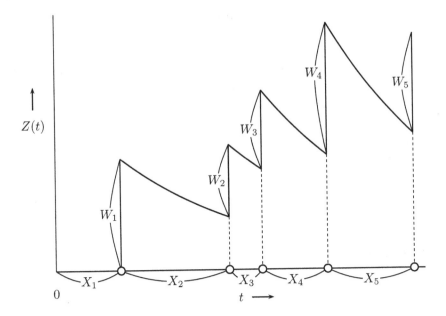

Fig. 2.3. Process for a cumulative damage model with annealing

(3) Damage with Annealing

The total damage in the usual reliability models is additive and does not decrease. In some materials, annealing, *i.e.*, lessening the damage, can take place such as rubber, fiber reinforced plastics, and polyurethane. We show two examples, using the results of [83].

Takács considered the following damage model: If a unit suffers damage W due to shock then its damage after time duration t is reduced to $We^{-\alpha t}$ $(0 < \alpha < \infty)$. Define

$$Z(t) \equiv \sum_{j=1}^{N(t)} W_j \exp[-\alpha(t - S_j)], \tag{2.54}$$

where $S_j \equiv \sum_{i=1}^{j} X_i$ $(j = 1, 2, \dots)$ (Figure 2.3). This also corresponds to the shot noise model in **(2)** of Section 10.1.

Suppose that shocks occur in a Poisson process with rate λ. Then, $\Phi(t, x) \equiv \Pr\{Z(t) \le x\}$ forms the following renewal equation [83, p. 105]:

$$\frac{\partial \Phi(t, x)}{\partial t} = -\lambda \left\{ \Phi(t, x) - \int_0^x G[(x - y)e^{-\alpha t}] \, dy \Phi(t, y) \right\}, \tag{2.55}$$

and its LS transform is

$$\frac{\partial \Phi^*(t, s)}{\partial t} = -\lambda[1 - G^*(se^{-\alpha t})]\Phi^*(t, s), \tag{2.56}$$

where $\Phi^*(t, s) \equiv \int_0^\infty e^{-sx} d\Phi(t, x)$ and $G^*(s) \equiv \int_0^\infty e^{-sx} dG(x)$. Solving this differential equation,

$$\Phi^*(t, s) = \exp\left\{-\lambda \int_0^t [1 - G^*(se^{-\alpha u})]\, du\right\}, \tag{2.57}$$

$$E\{Z(t)\} = -\left.\frac{\partial \Phi^*(t, s)}{\partial s}\right|_{s=0} = \frac{\lambda(1 - e^{-\alpha t})}{\alpha\mu}. \tag{2.58}$$

In addition, if $1/\mu = E\{W_j\} < \infty$, then $\lim_{t\to\infty} \Pr\{Z(t) \leq x\}$ exists and its LS transform is

$$\Phi^*(\infty, s) = \exp\left[-\frac{\lambda}{\alpha} \int_0^1 \frac{1 - G^*(su)}{u}\, du\right]. \tag{2.59}$$

Example 2.4.
(i) When $G(x) = 1 - e^{-\mu x}$,

$$\Phi^*(t, s) = \left(\frac{s + \mu e^{\alpha t}}{s + \mu}\right)^\nu e^{-\lambda t},$$

where $\nu \equiv \lambda/\alpha$. Thus, by inversion,

$$\Pr\{Z(t) \leq x\} = e^{-\lambda t} \sum_{j=0}^\infty \binom{\nu + j - 1}{j}(1 - e^{-\alpha t})^j \sum_{i=j}^\infty \frac{(\mu x e^{\alpha t})^i}{i!} \exp(-\mu x e^{\alpha t}).$$

In a similar way,

$$\Phi^*(\infty, s) = \left(\frac{\mu}{s + \mu}\right)^\nu,$$

$$\lim_{t\to\infty} \Pr\{Z(t) \leq x\} = \int_0^x \frac{\mu(\mu u)^{\nu-1}}{\Gamma(\nu)} e^{-\mu u}\, du,$$

that is a gamma distribution with mean ν/μ.
(ii) When $G(x) \equiv 0$ for $x < 1/\mu$ and 1 for $x \geq 1/\mu$, i.e., the damage due to each shock is constant and its amount is $1/\mu$. From the results [83, p. 129],

$$\Phi^*(\infty, s) = \left(\frac{\mu}{s\gamma}\right)^\nu \exp\left(-\nu \int_{1/\mu}^\infty \frac{e^{-su}}{u}\, du\right),$$

where $\gamma \equiv e^C = 1.781072\cdots$ and $C \equiv 0.577215\cdots$ that is Euler's constant. By inversion,

$$\lim_{t\to\infty} \Pr\{Z(t) \leq x\} = \frac{x^\nu + \sum_{j=1}^\infty [(-1)^j \nu^j/j!] \int_{j/\mu}^x (x - u)^\nu I^{(j)}(u)\, du}{(\gamma/\mu)^\nu \Gamma(1 + \nu)},$$

where $I(y)$ is uniform over $[0, 1/\mu]$. ∎

(4) n Different Units

Consider a system with n different units that are independent of each other. Successive shocks occur at time interval X_j with distribution $F(t) \equiv \Pr\{X_j \leq t\}$ $(j = 1, 2, \dots)$. Each shock causes some damage to unit i $(i = 1, 2, \dots, n)$ in the amount $W_{i;j}$ with distribution $G_i(x) \equiv \Pr\{W_{i;j} \leq x\}$ for all $j \geq 1$, where $W_{i;j}$ might be zero. Each unit fails when its total damage has exceeded its failure level K_i $(i = 1, 2, \dots, n)$. A series system with n units subject to shocks was considered [137].

One typical example of this model would be the damage to railroad tracks, ties and pantographs. Such damage is mainly due to the number and sizes of running trains and depends on the weight and the speed of trains. In the case of $n = 3$, X_j is the time interval of trains, and $W_{i;j}$ $(i = 1, 2, 3)$ are the amounts of damage to the railroad tracks, ties, and pantographs, respectively, produced by one running train.

Letting $Z_i(t)$ denote the total damage to unit i $(i = 1, 2, \dots, n)$ at time t, the joint distribution of $Z_i(t)$ is

$$\Pr\{Z_i(t) \leq x_i \ (i = 1, 2, \dots, n)\}$$
$$= \sum_{j=0}^{\infty} \Pr\{Z_i(t) \leq x_i \ (i = 1, 2, \dots, n) | N(t) = j\} \Pr\{N(t) = j\}. \quad (2.60)$$

From the assumption that each amount of damage occurs independently,

$$\Pr\{Z_i(t) \leq x_i \ (i = 1, 2, \dots, n) | N(t) = j\} = \prod_{i=1}^{n} G_i^{(j)}(x_i).$$

Thus, the joint distribution is

$$\Pr\{Z_i(t) \leq x_i \ (i = 1, 2, \dots, n)\} = \sum_{j=0}^{\infty} \left[\prod_{i=1}^{n} G_i^{(j)}(x_i) \right] [F^{(j)}(t) - F^{(j+1)}(t)].$$
$$(2.61)$$

Suppose that a system fails when at least one of n units exceeds a failure level K_i, i.e., the system is a n-unit series system. Then, the first-passage time distribution to system failure is

$$\Pr\{Y \leq t\} = 1 - \Pr\{Z_i(t) \leq K_i \ (i = 1, 2, \dots, n)\}$$
$$= \sum_{j=0}^{\infty} \left[1 - \prod_{i=1}^{n} G_i^{(j)}(K_i) \right] [F^{(j)}(t) - F^{(j+1)}(t)], \quad (2.62)$$

and its mean time is

$$E\{Y\} = \frac{1}{\lambda} \sum_{j=0}^{\infty} \left[\prod_{i=1}^{n} G_i^{(j)}(K_i) \right]. \quad (2.63)$$

Next, when a system fails if all of n units exceed a failure level K_i, *i.e.,* the system is an n-unit parallel system, the first-passage time distribution to system failure is

$$\Pr\{Y \le t\} = \sum_{j=0}^{\infty} \left\{ \prod_{i=1}^{n} [1 - G_i^{(j)}(K_i)] \right\} [F^{(j)}(t) - F^{(j+1)}(t)], \qquad (2.64)$$

and its mean time is

$$E\{Y\} = \frac{1}{\lambda} \sum_{j=0}^{\infty} \left\{ 1 - \prod_{i=1}^{n} [1 - G_i^{(j)}(K_i)] \right\}. \qquad (2.65)$$

When shocks occur in a nonhomogeneous Poisson process with a mean value function $H(t)$, the first-passage time distributions and their mean times are derived by replacing $F^{(j)}(t) - F^{(j+1)}(t)$ with $\{[H(t)]^j/j!\}e^{-H(t)}$ formally.

Furthermore, suppose that a shock does no damage to unit i with probability $q_i \equiv 1 - p_i$, and otherwise, does some positive damage $W_{i;j}$ with distribution $G_i(x)$. In this case,

$$\Pr\{Z_i(t) \le x_i \ (i = 1, 2, \ldots, n) | N(t) = j\} = \prod_{i=1}^{n} \left[\sum_{m=0}^{j} \binom{j}{m} q_i^m p_i^{j-m} G_i^{(j-m)}(x_i) \right],$$
$$(2.66)$$

and hence, we can get the first-passage time distributions and their mean times from (2.62)–(2.65).

Example 2.5. Suppose that any amount of damage to unit i incurred from shocks is constant $1/\mu_i$, *i.e.,* $G_i(x) = 0$ for $x < 1/\mu_i$ and 1 for $x \ge 1/\mu_i$. Let $K_m \equiv \min\{\mu_1 K_1, \mu_2 K_2, \ldots, \mu_n K_n\}$ and $K_M \equiv \max\{\mu_1 K_1, \mu_2 K_2, \ldots, \mu_n K_n\}$. The first-passage time distribution and its mean time for a series system are, from (2.62) and (2.63),

$$\Pr\{Y \le t\} = F^{([K_m]+1)}(t), \qquad E\{Y\} = \frac{1}{\lambda}([K_m] + 1),$$

and for a parallel system are, from (2.64) and (2.65),

$$\Pr\{Y \le t\} = F^{([K_M]+1)}(t), \qquad E\{Y\} = \frac{1}{\lambda}([K_M] + 1),$$

where $[x]$ denotes the greatest integer contained in x.

Moreover, when $F(t) = 1 - e^{-\lambda t}$ and $K_m \ge 1$, the failure rate is, for a series system,

$$r(t) = \frac{\lambda(\lambda t)^{[K_m]}/[K_m]!}{\sum_{j=0}^{[K_m]}(\lambda t)^j/j!},$$

and for a parallel system,

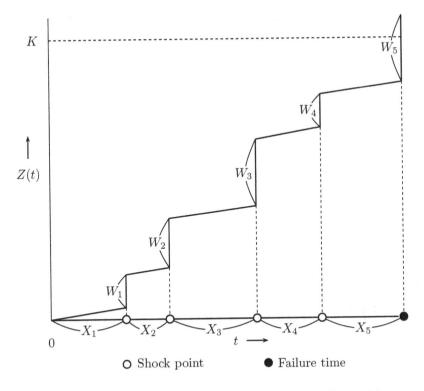

Fig. 2.4. Process for a cumulative damage model with two kinds of damages

$$r(t) = \frac{\lambda(\lambda t)^{[K_M]}/[K_M]!}{\sum_{j=0}^{[K_M]}(\lambda t)^j/j!},$$

both of which are $r(0) = 0$, and increase monotonically and become $r(\infty) = \lambda$ that is the constant failure rate of an exponential distribution $(1 - e^{-\lambda t})$. If $K_M < 1$, then $r(t) = \lambda$ for all $t \geq 0$. ∎

(5) Increasing Damage with Time

Consider the cumulative damage model with two kinds of damage (see Figure 2.4). One of them is caused by shock and is additive, and the other increases proportionately with time, that is, the total damage is accumulated subject to shocks and time at the rate of constant α ($\alpha > 0$), independent of shocks. A unit fails whether the total damage is exceeded with time or has exceeded a failure level K at some shock, and its failure is detected only at the time of shocks. Such a model would be the life of dry and storage batteries. A battery supplies electric power that is stored by chemical change according to its need. However, oxidation and deoxidation always occur irrespective of its use, that is, a battery always discharges a small quantity of electricity with time, and finally, it cannot be used.

Suppose that $S_j \equiv X_1 + X_2 + \cdots + X_j$, $Z_j \equiv W_1 + W_2 + \cdots + W_j$ ($j = 1, 2, \ldots$), and $S_0 \equiv Z_0 \equiv 0$. Because $\Pr\{S_j \leq t\} = F^{(j)}(t)$ where $\Pr\{Z_j \leq x\} = G^{(j)}(x)$ ($j = 0, 1, 2, \ldots$), the distribution of time to detect a failure at some shock is

$$\Pr\{Y \leq t\} = \sum_{j=0}^{\infty} \Pr\{Z_j + \alpha S_j < K \leq Z_{j+1} + \alpha S_{j+1}, S_{j+1} \leq t\}$$

$$= \sum_{j=0}^{\infty} \int_0^t \left\{ \int_0^{t-u} [G^{(j)}(K - \alpha u) - G^{(j+1)}(K - \alpha(u + x))] \, dF(x) \right\} dF^{(j)}(u),$$

(2.67)

where note that $G^{(j)}(x) \equiv 0$ for $x < 0$. Thus, the mean time to detect a failure at some shock is

$$E\{Y\} =$$

$$\sum_{j=0}^{\infty} \int_0^{\infty} \left\{ \int_0^{\infty} (t + x)[G^{(j)}(K - \alpha t) - G^{(j+1)}(K - \alpha(t + x))] \, dF(x) \right\} dF^{(j)}(t)$$

$$= \frac{1}{\lambda} \sum_{j=0}^{\infty} \int_0^{K/\alpha} G^{(j)}(K - \alpha t) \, dF^{(j)}(t).$$

(2.68)

Similarly, the probability that the failure is detected at the $(j + 1)$th shock is

$$p_{j+1} = \int_0^{\infty} \left\{ \int_0^{\infty} [G^{(j)}(K - \alpha t) - G^{(j+1)}(K - \alpha(t + x))] \, dF(x) \right\} dF^{(j)}(t)$$

$$= \int_0^{K/\alpha} G^{(j)}(K - \alpha t) \, dF^{(j)}(t) - \int_0^{K/\alpha} G^{(j+1)}(K - \alpha t) \, dF^{(j+1)}(t)$$

$$(j = 0, 1, 2, \ldots), \qquad (2.69)$$

and the failure rate is

$$r_{j+1} = \frac{\int_0^{K/\alpha} G^{(j)}(K - \alpha t) \, dF^{(j)}(t) - \int_0^{K/\alpha} G^{(j+1)}(K - \alpha t) \, dF^{(j+1)}(t)}{\int_0^{K/\alpha} G^{(j)}(K - \alpha t) \, dF^{(j)}(t)}$$

$$(j = 0, 1, 2, \ldots). \qquad (2.70)$$

This corresponds to the model where a failure level $K(t)$ at time t decreases with time t, i.e., $K(t) = K - \alpha t$.

Example 2.6. It is intuitively estimated from (2.11) that because the average damage per unit of time is $\alpha + \lambda/\mu$, the mean time until the total damage has exceeded a failure level K is approximately

Table 2.1. Mean time to failure for two kinds of damage when $1/\lambda = 1$

$\alpha\mu$	$\mu K = 1$		$\mu K = 5$		$\mu K = 10$	
	λl	$\lambda E\{Y\}$	λl	$\lambda E\{Y\}$	λl	$\lambda E\{Y\}$
0.0	2.0	2.000	6.0	6.000	11.0	11.000
0.2	1.8	1.705	5.2	5.078	9.3	9.294
0.4	1.7	1.521	4.6	4.392	8.1	7.989
0.6	1.6	1.410	4.1	3.907	7.3	7.049
0.8	1.6	1.334	3.8	3.543	6.6	6.333
1.0	1.5	1.286	3.5	3.260	6.0	5.770
2.0	1.3	1.162	2.7	2.450	4.3	4.121
4.0	1.2	1.086	2.0	1.843	3.0	2.845

$$ l = \frac{1}{\lambda} \left(\frac{K}{\alpha/\lambda + 1/\mu} + 1 \right). $$

Table 2.1 presents $\lambda E\{Y\}$ and λl for $\alpha\mu$ and μK when $F(t) = 1 - e^{-\lambda t}$, $G(x) = 1 - e^{-\mu x}$, and $1/\lambda = 1$. When $\alpha = 0$, this corresponds to the standard cumulative model given in Example 2.2. This table indicates that l shows a good upper bound for the mean time to failure. In actual models, l would be easily computed, and it would be used practically as one estimation of their mean failure times. ∎

Finally, if the total damage increases exponentially, *i.e.*,

$$ Z(t) = \sum_{j=1}^{N(t)} W_j \exp\left[\alpha(t - S_j)\right], \tag{2.71} $$

then by arguments similar to those of **(3)**, when $F(t) = 1 - e^{-\lambda t}$,

$$ \Phi^*(t, s) = \exp\left\{ -\lambda \int_0^t [1 - G^*(se^{\alpha u})]\, du \right\}, \tag{2.72} $$

$$ E\{Z(t)\} = \frac{\lambda(e^{\alpha t} - 1)}{\alpha\mu}, \tag{2.73} $$

$$ \Phi^*(\infty, s) = \exp\left[-\frac{\lambda}{\alpha} \int_1^\infty \frac{1 - G^*(su)}{u}\, du \right]. \tag{2.74} $$

This corresponds to the model where the total damage due to shocks is additive and also increases exponentially with time.

3

Basic Replacement Policies

Consider a unit that should operate over an infinite time span. It is assumed that shocks occur in random times and each shock causes a random amount of damage to a unit. These damages are additive, and a unit fails when the total damage has exceeded a failure level K. When the failure during actual operation is costly or dangerous, it is of great importance to avoid such terrible situations. It would be wise to exchange a unit at a lower cost before its failure. The replacement after failure and before failure is called *corrective replacement* and *preventive replacement*, respectively. We may consider *damage* as *cost* incurred from shocks. In this case, this corresponds to the maintenance model where a unit is replaced when the total cost incurred for some maintenance has exceeded a threshold level K.

This is the maintenance model for a single unit, where its failure is very serious, and sometimes may incur a heavy loss. If we have no information on the condition of a unit, its maintenance should be done at planned times. On the other hand, if we could get the number of shocks up to now and the amount of damage at shock times or at inspection times, its maintenance should be done at a prespecified number of shocks or at a damage level before failure, respectively.

Suppose that a unit is replaced with a new one at failure. It may be wise to do some maintenance at a lower cost before failure. The optimum control-limit policies where a unit is replaced at a threshold level was derived, when it fails with a known probability that is a function of the total damage [42–45]. More discussions on such replacement policies were carried out [138–146]. Such replacements were summarized [147, 148]. On the other hand, the replacement models where a unit is replaced at a planned time T were proposed [46–50]. Furthermore, the cumulative damage model where the total damage is decreasing at a known restoration rate was proposed [149–152]. Recently, a variety of replacement models subject to shocks were studied [153–160]. Replacement policies for multistate degraded systems subject to random shocks were discussed [161–165]. A δ-shock model, where the second shock will cause the failure if the time interval between two successive shocks is less than δ, was proposed [166, 167].

This chapter is written based on [51, 168] and adds some new results by combining the theories of cumulative processes [11] and maintenance [1]. In Section 3.1, a unit is replaced before failure at a planned time T, at a shock number N, or at a damage level Z, whichever occurs first. Introducing the respective replacement costs for T, N, and Z, we obtain the expected cost rates. In Section 3.2, we derive analytically optimum policies that minimize the expected cost rates for the three policies. Some optimum policies are compared with other values in numerical examples. In Section 3.3, we propose five modified replacement models that would be useful in practical fields and give more interesting research topics for further study.

3.1 Three Replacement Policies

Suppose that a unit begins to operate at time 0 and its damage level is 0. Let $N(t)$ be the number of shocks in time t. It is assumed that the probability that j shocks occur in $[0, t]$ is $F_j(t)$ $(j = 1, 2, \cdots)$, where $F_0(t) \equiv 1$, *i.e.*, the probability that j shocks occur exactly in $[0, t]$ is

$$\Pr\{N(t) = j\} = F_j(t) - F_{j+1}(t) \qquad (j = 0, 1, 2, \cdots).$$

An amount W_j of damage due to the jth shock has an identical distribution $G(x) \equiv \Pr\{W_j \le x\}$ with finite mean $1/\mu$, where $\overline{G}(x) \equiv 1 - G(x)$ and $1/\mu \equiv \int_0^\infty \overline{G}(x)\mathrm{d}x < \infty$. Furthermore, the total damage is additive, and its level is investigated and is known only at shock times. The unit fails when the total damage has exceeded a failure level K at some shock, its failure is immediately detected, and it is replaced with a new one.

As the preventive replacement policy, the unit is replaced before failure at a planned time T $(0 < T \le \infty)$, at a shock number N $(N = 1, 2, \cdots)$, or at a damage level Z $(0 \le Z \le K)$, whichever occurs first. In addition, it is assumed that the unit is replaced at K or Z without replacing it at N, respectively, when the total damage has exceeded K or Z at shock N.

The probability that the unit is replaced at time T is

$$P_T = \sum_{j=0}^{N-1} [F_j(T) - F_{j+1}(T)]G^{(j)}(Z), \qquad (3.1)$$

the probability that it is replaced at shock N is

$$P_N = F_N(T)G^{(N)}(Z), \qquad (3.2)$$

the probability that it is replaced at damage Z is

$$P_Z = \sum_{j=0}^{N-1} F_{j+1}(T) \int_0^Z [G(K - x) - G(Z - x)]\, \mathrm{d}G^{(j)}(x), \qquad (3.3)$$

and the probability that it is replaced at failure level K, *i.e.*, corrective replacement is done, is

$$P_K = \sum_{j=0}^{N-1} F_{j+1}(T) \int_0^Z \overline{G}(K-x)\, \mathrm{d}G^{(j)}(x), \tag{3.4}$$

where $\varphi^{(j)}(x)$ $(j = 1, 2, \cdots)$ denotes the j-fold Stieltjes convolution of any distribution $\varphi(x)$ with itself and $\varphi^{(0)}(x) \equiv 1$ for $x \geq 0$. It is clearly shown that $P_T + P_N + P_Z + P_K = 1$. Similarly, the mean time to replacement is

$$T \sum_{j=0}^{N-1} [F_j(T) - F_{j+1}(T)]G^{(j)}(Z) + G^{(N)}(Z) \int_0^T t\, \mathrm{d}F_N(t)$$

$$+ \sum_{j=0}^{N-1} \int_0^T t\, \mathrm{d}F_{j+1}(t) \int_0^Z [G(K-x) - G(Z-x)]\, \mathrm{d}G^{(j)}(x)$$

$$+ \sum_{j=0}^{N-1} \int_0^T t\, \mathrm{d}F_{j+1}(t) \int_0^Z \overline{G}(K-x)\, \mathrm{d}G^{(j)}(x)$$

$$= \sum_{j=0}^{N-1} G^{(j)}(Z) \int_0^T [F_j(t) - F_{j+1}(t)]\, \mathrm{d}t. \tag{3.5}$$

For the above replacement model, we introduce the following replacement costs: Cost c_T is incurred for replacement at time T, and c_N, c_Z, and c_K are the respective replacement cost at shock N, damage Z, and failure level K, where cost c_K is higher than the three costs c_T, c_N, and c_Z. Then, the total expected cost until replacement, given that the unit began to operate at time 0, is

$$\widehat{C}(T, N, Z) = c_T P_T + c_N P_N + c_Z P_Z + c_K P_K$$

$$= c_K - (c_K - c_T) \sum_{j=0}^{N-1} [F_j(T) - F_{j+1}(T)]G^{(j)}(Z)$$

$$- (c_K - c_N)F_N(T)G^{(N)}(Z)$$

$$- (c_K - c_Z) \sum_{j=0}^{N-1} F_{j+1}(T) \int_0^Z [G(K-x) - G(Z-x)]\, \mathrm{d}G^{(j)}(x). \tag{3.6}$$

We call the time interval from one replacement to the next replacement one cycle. Then, the pairs of time and cost in each cycle are independently and identically distributed, and both have finite means. Thus, from (1.34) in a renewal reward process, the expected cost per unit of time for an infinite interval is

$$C(T, N, Z) = \frac{\text{Expected cost of one cycle}}{\text{Mean time of one cycle}}, \tag{3.7}$$

that is called the *expected cost rate*. Thus, dividing (3.6) by (3.5),

$$C(T, N, Z) = \frac{\begin{aligned} &c_K - (c_K - c_T)\sum_{j=0}^{N-1}[F_j(T) - F_{j+1}(T)]G^{(j)}(Z) \\ &- (c_K - c_N)F_N(T)G^{(N)}(Z) \\ &- (c_K - c_Z)\sum_{j=0}^{N-1} F_{j+1}(T)\int_0^Z [G(K-x) - G(Z-x)]\,\mathrm{d}G^{(j)}(x) \end{aligned}}{\sum_{j=0}^{N-1} G^{(j)}(Z)\int_0^T [F_j(t) - F_{j+1}(t)]\,\mathrm{d}t}. \tag{3.8}$$

When the unit is replaced only after failure, the expected cost rate is

$$C \equiv \lim_{\substack{T \to \infty \\ N \to \infty \\ Z \to K}} C(T, N, Z)$$

$$= \frac{c_K}{\sum_{j=0}^{\infty} G^{(j)}(K)\int_0^{\infty}[F_j(t) - F_{j+1}(t)]\,\mathrm{d}t}. \tag{3.9}$$

Furthermore, denoting c_k as the mean time for replacement at k ($k = T, N, Z, K$), the availability $A(T, N, Z)$ ((2.24) of [1]) is

$$A(T, N, Z) \equiv \frac{\text{Mean time to replacement}}{\text{Mean time to replacement} + \text{Mean time for replacement}}$$

$$= 1 \Big/ \left\{ 1 + \frac{c_T P_T + c_N P_N + c_Z P_Z + c_K P_K}{\sum_{j=0}^{N-1} G^{(j)}(Z)\int_0^T [F_j(t) - F_{j+1}(t)]\,\mathrm{d}t} \right\}. \tag{3.10}$$

Thus, the policy maximizing $A(T, N, Z)$ is theoretically the same as minimizing the expected cost rate $C(T, N, Z)$ in (3.8).

3.2 Optimum Policies

We discuss analytically an optimum planned time T^*, shock number N^*, and damage level Z^* that minimize the expected cost rates when $F_j(t) \equiv F^{(j)}(t)$ ($j = 1, 2, \cdots$), *i.e.*, shocks occur in a renewal process with a general distribution $F(t)$ and its finite mean $1/\lambda$.

(1) Optimum T^*

Suppose that a unit is replaced at time T ($0 < T \leq \infty$) or at failure, whichever occurs first. Then, the expected cost rate is, from (3.8),

$$C_1(T) \equiv \lim_{\substack{N \to \infty \\ Z \to K}} C(T, N, Z)$$

$$= \frac{c_K - (c_K - c_T)\sum_{j=0}^{\infty}[F^{(j)}(T) - F^{(j+1)}(T)]G^{(j)}(K)}{\sum_{j=0}^{\infty} G^{(j)}(K)\int_0^T [F^{(j)}(t) - F^{(j+1)}(t)]\,\mathrm{d}t}. \tag{3.11}$$

It can be easily seen that $\lim_{T \to 0} C_1(T) = \infty$, and from (3.9),

$$C_1 \equiv \lim_{T \to \infty} C_1(T) = \frac{c_K}{[1 + M_G(K)]/\lambda},\qquad(3.12)$$

where $M_G(K) \equiv \sum_{j=1}^{\infty} G^{(j)}(K)$, and note that the denominator of the right-hand side represents the mean time to failure given in (2.11). Thus, there exists a positive T^* $(0 < T^* \le \infty)$ that minimizes $C_1(T)$.

We seek an optimum time T^* that minimizes $C_1(T)$ in (3.11) for $c_K > c_T$. Let $f(t)$ be a density function of $F(t)$, $f^{(j)}(t)$ $(j = 1, 2, \cdots)$ be the j-fold Stieltjes convolution of $f(t)$ with itself, and $f^{(0)}(t) \equiv 0$ for $t \ge 0$. Then, differentiating $C_1(T)$ with respect to T and setting it equal to zero,

$$Q(T) \sum_{j=0}^{\infty} G^{(j)}(K) \int_0^T [F^{(j)}(t) - F^{(j+1)}(t)]\, dt$$

$$- \sum_{j=0}^{\infty} F^{(j+1)}(T)[G^{(j)}(K) - G^{(j+1)}(K)] = \frac{c_T}{c_K - c_T},\qquad(3.13)$$

where

$$Q(T) \equiv \frac{\sum_{j=0}^{\infty} f^{(j+1)}(T)[G^{(j)}(K) - G^{(j+1)}(K)]}{\sum_{j=0}^{\infty} [F^{(j)}(T) - F^{(j+1)}(T)]G^{(j)}(K)}.$$

It can be clearly seen that if $Q(T)$ is strictly increasing in T, then the left-hand side of (3.13) is also strictly increasing from 0 to $Q(\infty)(1/\lambda)[1 + M_G(K)] - 1$, where $Q(\infty) \equiv \lim_{T \to \infty} Q(T)$. Thus, if $Q(\infty)[1 + M_G(K)] > \lambda c_K/(c_K - c_T)$, then there exists a finite and unique T^* that satisfies (3.13), and the resulting cost rate is

$$C_1(T^*) = (c_K - c_T)Q(T^*).\qquad(3.14)$$

Conversely, if $Q(\infty)[1 + M_G(K)] \le \lambda c_K/(c_K - c_T)$, then $T^* = \infty$, i.e., the unit is replaced only at failure, and the expected cost rate is given in (3.12).

If a failure level K is distributed according to a general distribution $L(x)$ as shown in (2) of Section 2.5, the expected cost rate becomes

$$C_1(T) = \frac{c_K - (c_K - c_T)\sum_{j=0}^{\infty}[F_j(T) - F_{j+1}(T)]\int_0^{\infty} G^{(j)}(x)\, dL(x)}{\sum_{j=0}^{\infty} \int_0^T [F_j(t) - F_{j+1}(t)]\, dt \int_0^{\infty} G^{(j)}(x)\, dL(x)}.\qquad(3.15)$$

In particular, suppose that shocks occur in a nonhomogeneous Poisson process and a failure level K is distributed exponentially, i.e., $F_j(t) = \sum_{i=j}^{\infty}\{[H(t)]^j/j!\}$ $\times e^{-H(t)}$ $(j = 0, 1, 2, \cdots)$ and $L(x) = 1 - e^{-\theta x}$. Then, the expected cost rate is rewritten as

$$C_1(T) = \frac{c_K - (c_K - c_T)e^{-[1-G^*(\theta)]H(T)}}{\int_0^T e^{-[1-G^*(\theta)]H(t)}\, dt},\qquad(3.16)$$

where $G^*(\theta)$ denotes the Laplace–Stieltjes transform of $G(x)$, i.e., $G^*(\theta) \equiv \int_0^{\infty} e^{-\theta x} dG(x)$ for $\theta > 0$.

We seek an optimum time T^* that minimizes $C_1(T)$ in (3.16). First, it is easily noted that the problem of minimizing $C_1(T)$ is the same standard age replacement problem with a failure distribution $(1 - \exp\{-[1 - G^*(\theta)]H(t)\})$ in Chapter 3 of [1]. Let $h(t)$ be an intensity function of a nonhomogeneous Poisson process, i.e., $h(t) \equiv dH(t)/dt$ and $H(t) = \int_0^t h(u)du$. Then, differentiating $C_1(T)$ with respect to T and setting it equal to zero,

$$[1 - G^*(\theta)]h(T) \int_0^T e^{-[1-G^*(\theta)]H(t)}\, dt + e^{-[1-G^*(\theta)]H(T)} = \frac{c_K}{c_K - c_T}. \quad (3.17)$$

Letting $Q_1(T)$ denote the left-hand side of (3.17), it can be easily seen that if $h(t)$ is strictly increasing, then $Q_1(T)$ is also strictly increasing from 1 to

$$Q_1(\infty) \equiv \lim_{T \to \infty} Q_1(T) = [1 - G^*(\theta)]h(\infty) \int_0^\infty e^{-[1-G^*(\theta)]H(t)}\, dt.$$

Therefore, we have the following optimum policy:

(i) If $h(t)$ is strictly increasing and $Q_1(\infty) > c_K/(c_K - c_T)$, then there exists a finite and unique T^* $(0 < T^* < \infty)$ that satisfies (3.17), and the resulting cost rate is

$$C_1(T^*) = (c_K - c_T)[1 - G^*(\theta)]h(T^*). \quad (3.18)$$

(ii) If $h(t)$ is strictly increasing and $Q_1(\infty) \leq c_K/(c_K - c_T)$ or $h(t)$ is nonincreasing, then $T^* = \infty$, and the expected cost rate is

$$C_1(\infty) \equiv \lim_{T \to \infty} C_1(T) = \frac{c_K}{\int_0^\infty e^{-[1-G^*(\theta)]H(t)}\, dt}. \quad (3.19)$$

In the case of (ii), it is of interest that there does not exist any finite time T^* to minimize $C_1(T)$ when shocks occur in a Poisson process, i.e., $h(t) = \lambda$.

(2) Optimum N^*

Suppose that a unit is replaced at shock N $(N = 1, 2, \cdots)$ or at failure, whichever occurs first. Then, the expected cost rate is, from (3.8),

$$
\begin{aligned}
C_2(N) &\equiv \lim_{\substack{T \to \infty \\ Z \to K}} C(T, N, Z) \\
&= \frac{c_K - (c_K - c_N)G^{(N)}(K)}{(1/\lambda)\sum_{j=0}^{N-1} G^{(j)}(K)} \quad (N = 1, 2, \cdots).
\end{aligned} \quad (3.20)
$$

In particular, when $N = 1$, i.e., the unit is always replaced at the first shock, the expected cost rate is

$$C_2(1) = \lambda[c_K - (c_K - c_N)G(K)]. \quad (3.21)$$

Forming the inequality $C_2(N+1) - C_2(N) \geq 0$ to seek an optimum number N^* that minimizes $C_2(N)$ for $c_K > c_N$,

$$Q_2(N+1) \sum_{j=0}^{N-1} G^{(j)}(K) - [1 - G^{(N)}(K)] \geq \frac{c_N}{c_K - c_N} \qquad (N = 1, 2, \cdots), \quad (3.22)$$

where

$$Q_2(N) \equiv \frac{G^{(N-1)}(K) - G^{(N)}(K)}{G^{(N-1)}(K)} \qquad (N = 1, 2, \cdots).$$

If $Q_2(N)$ is strictly increasing in N, i.e., $G^{(j+1)}(x)/G^{(j)}(x)$ is strictly decreasing in j, then the left-hand side of (3.22) is also strictly increasing in N to $Q_2(\infty)[1 + M_G(K)] - 1$, where $Q_2(\infty) \equiv \lim_{N \to \infty} Q_2(N) \leq 1$. Thus, if $Q_2(\infty)[1 + M_G(K)] > c_K/(c_K - c_N)$, then there exists a finite and unique minimum number N^* $(1 \leq N^* < \infty)$ that satisfies (3.22), and the expected cost rate is

$$\lambda(c_K - c_N)Q_2(N^*) < C_2(N^*) \leq \lambda(c_K - c_N)Q_2(N^* + 1). \qquad (3.23)$$

Conversely, if $Q_2(\infty)[1 + M_G(K)] \leq c_K/(c_K - c_N)$, then $N^* = \infty$. Note that $Q_2(N)$ corresponds to the discrete failure rate r_N given in (2.15), and $Q_2(N+1)$ represents the probability that the unit surviving at the Nth shock will fail at the $(N+1)$th shock. In general, $Q_2(N)$ would increase to 1. In this case, if $M_G(K) > c_N/(c_K - c_N)$, i.e., the expected number of shocks before failure is greater than $c_N/(c_K - c_N)$, then a finite N^* exists uniquely.

(3) Optimum Z^*

Suppose that a unit is replaced at damage Z $(0 \leq Z \leq K)$ or at failure, whichever occurs first. Then, the expected cost rate is, from (3.8),

$$C_3(Z) \equiv \lim_{\substack{T \to \infty \\ N \to \infty}} C(T, N, Z)$$

$$= \frac{c_K - (c_K - c_Z)[G(K) - \int_0^Z \overline{G}(K - x) \, \mathrm{d}M_G(x)]}{[1 + M_G(Z)]/\lambda}. \qquad (3.24)$$

When $Z = 0$, $C_3(0)$ agrees with $C_2(1)$ in (3.21) when $c_Z = c_N$.

We seek an optimum level Z^* that minimizes $C_3(Z)$ in (3.24) for $c_K > c_Z$. Differentiating $C_3(Z)$ with respect to Z and setting it equal to zero,

$$\int_{K-Z}^{K} [1 + M_G(K - x)] \, \mathrm{d}G(x) = \frac{c_Z}{c_K - c_Z}. \qquad (3.25)$$

The left-hand side of (3.25) is strictly increasing from 0 to $M_G(K)$. Thus, if $M_G(K) > c_Z/(c_K - c_Z)$, then there exists a finite and unique Z^* $(0 < Z^* < K)$ that satisfies (3.25), and its resulting cost rate is

$$C_3(Z^*) = \lambda(c_K - c_Z)\overline{G}(K - Z^*). \tag{3.26}$$

Conversely, if $M_G(K) \le c_Z/(c_K - c_Z)$, then $Z^* = K$, i.e., the unit should be replaced only at failure, and the expected cost rate is given in (3.12).

If $G(x)$ has an IFR property, then from (1.20), $\mu K \ge M_G(K) \ge \mu K - 1$, where $1/\mu \equiv E\{W_j\}$. Thus, if $\mu K > c_K/(c_K - c_Z)$, then an optimum Z^* $(0 < Z^* < K)$ exists uniquely, and if $\mu K \le c_Z/(c_K - c_Z)$, then $Z^* = K$. In addition, if the solutions Z_1 and Z_2 to satisfy

$$\int_{K-Z}^{K} [1 + \mu(K - x)]\, dG(x) = \frac{c_Z}{c_K - c_Z}, \tag{3.27}$$

and

$$\int_{K-Z}^{K} \mu(K - x)\, dG(x) = \frac{c_Z}{c_K - c_Z} \tag{3.28}$$

exist, respectively, then $Z_1 \le Z^* \le Z_2$.

Example 3.1.   Consider the replacement of car tires where the damage to the tire is a function of the running distance. If the running distance exceeds $K = 30,000$ km, the tire is regarded as failed and is not suitable for running. The distance traveled in one time unit is assumed to obey an exponential distribution with mean $1/\mu$, i.e., $G(x) = 1 - e^{-\mu x}$ and $M_G(x) = \mu x$. Then, cost c_Z represents the usual replacement cost of the tire and is 11,000 yen (about \$100). Cost c_K includes all costs resulting from the failure of tires in service, and will be higher than c_Z because there is a risk of accidents. From the above results, if $\mu K > c_Z/(c_K - c_Z)$, then there exists a finite and unique Z^* that satisfies

$$\mu Z e^{-\mu(K-Z)} = \frac{c_Z}{c_K - c_Z}.$$

Thus, we may replace the tire when the total running exceeds Z^* km before failure. In this case, the expected cost rate is $C_3(Z^*)/(\lambda c_Z) = 1/(\mu Z^*)$. On the other hand, if the tire is replaced only when the total distance has exceeded 30,000 km, then the expected cost is $C_3(K)/(\lambda c_Z) = (c_K/c_Z)/(1 + \mu K)$. Furthermore, from (3.28), Z_2 is given by the unique solutions of the following equations:

$$e^{-\mu K}\left[1 - (1 - \mu Z)e^{\mu Z}\right] = \frac{c_Z}{c_K - c_Z},$$

and $Z^* = Z_1 \le Z_2$.

Another simple method of replacement is to balance the ratio of replacement costs before and after failures against that of a damage level and a failure level, i.e.,

$$\frac{\tilde{Z}}{K + 1/\mu} = \frac{c_Z}{c_K}.$$

It is clearly seen that $Z^* > \tilde{Z}$ because $e^{\mu(K-Z)} > 1 + \mu(K - Z)$ for $K > Z$.

Table 3.1. Comparison of optimum damage level Z^* and approximate values Z_2 and $\widetilde{Z}$ for c_K/c_Z and $1/\mu$ when $K = 30,000$ km

$1/\mu$	$c_K/c_Z = 2$			$c_K/c_Z = 5$			$c_K/c_Z = 10$		
	Z^*	Z_2	$\widetilde{Z}$	Z^*	Z_2	$\widetilde{Z}$	Z^*	Z_2	$\widetilde{Z}$
100	29431	29431	15050	29293	29293	6020	29212	29212	3010
200	29004	29006	15100	28729	28730	6040	28568	28569	3020
300	28632	28635	15150	28220	28224	6060	27980	27983	3030
400	28296	28302	15200	27749	27755	6080	27429	27435	3040
500	27987	27996	15250	27306	27315	6100	26908	26917	3050
600	27700	27713	15300	26886	26900	6120	26410	26424	3060
700	27432	27449	15350	26486	26504	6140	25933	25952	3070
800	27179	27202	15400	26102	26127	6160	25473	25498	3080
900	26940	26970	15450	25734	25765	6180	25029	25061	3090
1000	26714	26751	15500	25379	25418	6200	24600	24639	3100

Table 3.1 presents the optimum value Z^*, upper value Z_2, and approximate value $\widetilde{Z}$ for $1/\mu$ and c_K/c_Z, that decrease with both $1/\mu$ and c_K/c_Z. This indicates that $\widetilde{Z} < Z^* \le Z_2$ shows a good approximation, however, $\widetilde{Z}$ is too small to compare with Z^*, so that the upper bound given in (3.28) would be very useful practically to compute an optimum policy when $G(x)$ and its mean $1/\mu$ are statistically estimated. ∎

Until now, it has been assumed that shocks occur in random times and their amount of damage is statistically estimated. Next, the amount of damage is investigated only through inspections that are made at periodic times, that is, the amount of damage is generated during $((j-1)t_0, jt_0]$ according to an identical distribution $G(x)$ for all j $(j = 1, 2, \cdots)$, and its total damage is known only at jt_0, i.e., at the end of each period. This corresponds to the damage model where shocks occur at a constant time t_0. Replacing $1/\lambda$ with t_0 in (3.24), we can obtain the expected cost rate and make a discussion similar to deriving an optimum policy.

3.3 Modified Replacement Models

This section considers some extended models of Section 3.1 in more general replacement forms and discusses optimum policies. Furthermore, we propose the combined preventive replacement models of planned time, shock number and damage level. These models would be more realistic than the basic ones, and moreover, offer interesting topics to reliability theoreticians.

(1) Modified Cost

The replacement costs may depend on the damage level at its replacement time. It is assumed that $c_0(x)$ $(0 \leq x \leq K)$ is an additional replacement cost that is variable for the total damage x with $c(0) = 0$, that is, cost $c_k + c_0(x)$ $(k = T, N, Z)$ is incurred for the replacement of the unit with damage x at time T, shock N, and damage Z, respectively, and cost $c_K + c_0(K)$ is incurred for the replacement at failure.

The expected cost when the unit is replaced at time T is

$$\sum_{j=0}^{N-1} [F_j(T) - F_{j+1}(T)] \int_0^Z [c_T + c_0(x)] \, dG^{(j)}(x), \qquad (3.29)$$

the expected cost when it is replaced at shock N is

$$F_N(T) \int_0^Z [c_N + c_0(x)] \, dG^{(N)}(x), \qquad (3.30)$$

and the expected cost when it is replaced at damage Z is

$$\sum_{j=0}^{N-1} F_{j+1}(T) \int_0^Z \int_{Z-x}^{K-x} [c_Z + c_0(x+y)] \, dG(y) \, dG^{(j)}(x). \qquad (3.31)$$

Thus, summing up (3.29)–(3.31), adding them to the replacement cost $[c_K + c_0(K)]P_K$, and dividing by (3.5), the expected cost rate is, from (3.7),

$$C(T, N, Z) = \frac{\begin{aligned} &c_K - (c_K - c_T)\sum_{j=0}^{N-1}[F_j(T) - F_{j+1}(T)]G^{(j)}(Z) \\ &\quad - (c_K - c_N)F_N(T)G^{(N)}(Z) \\ &\quad - (c_K - c_Z)\sum_{j=0}^{N-1}F_{j+1}(T)\int_0^Z [G(K-x) - G(Z-x)] \, dG^{(j)}(x) \\ &\quad + \sum_{j=0}^{N-1} F_{j+1}(T) \int_0^Z [\int_x^K \overline{G}(y-x) \, dc_0(y)] \, dG^{(j)}(x) \end{aligned}}{\sum_{j=0}^{N-1} G^{(j)}(Z) \int_0^T [F_j(t) - F_{j+1}(t)] \, dt}. \qquad (3.32)$$

It is difficult to discuss optimum policies analytically. In particular, it is assumed that shocks occur in a Poisson process with rate λ, the amount of damage due to each shock has an exponential distribution with mean $1/\mu$, and $c_0(x)$ is proportional to the total damage x, i.e., $F_j(t) = \sum_{i=j}^{\infty}[(\lambda t)^i/i!]e^{-\lambda t}$, $G^{(j)}(x) = \sum_{i=j}^{\infty}[(\mu x)^i/i!]e^{-\mu x}$, and $c_0(x) = c_0 x$.

The expected cost rate for the replacement at time T under the above conditions is

$$\begin{aligned} \frac{C_1(T)}{\lambda} &\equiv \lim_{\substack{N \to \infty \\ Z \to K}} \frac{C(T, N, Z)}{\lambda} \\ &= \frac{c_K - c_0/\mu - (c_K - c_T - c_0/\mu)\sum_{j=0}^{\infty}[F_j(T) - F_{j+1}(T)]G^{(j)}(K)}{\sum_{j=0}^{\infty} F_{j+1}(T)G^{(j)}(K)} \\ &\quad + \frac{c_0}{\mu}. \end{aligned} \qquad (3.33)$$

Differentiating $C_1(T)$ with respect to T and setting it equal to zero, for $c_K > c_T + c_0/\mu$,

$$Q(T)\sum_{j=0}^{\infty} F_{j+1}(T)G^{(j)}(K) - \sum_{j=0}^{\infty} \frac{(\lambda T)^j}{j!}e^{-\lambda T}[1 - G^{(j)}(K)] = \frac{c_T}{c_K - c_T - c_0/\mu},$$

(3.34)

where

$$Q(T) \equiv \frac{\sum_{j=0}^{\infty}[(\lambda T)^j/j!]e^{-\lambda T}[G^{(j)}(K) - G^{(j+1)}(K)]}{\sum_{j=0}^{\infty}[(\lambda T)^j/j!]e^{-\lambda T}G^{(j)}(K)}.$$

First, note that $[G^{(j)}(K) - G^{(j+1)}(K)]/G^{(j)}(K) = [(\mu K)^j/j!]/\sum_{i=j}^{\infty}[(\mu K)^i/i!]$ is strictly increasing from $e^{-\mu K}$ to 1 from Example 2.2. Next, when $[G^{(j)}(x) - G^{(j+1)}(x)]/G^{(j)}(x)$ is strictly increasing in j for any distribution $G(x)$, we can prove [131] that

$$Q(T) = \frac{\sum_{j=0}^{\infty}[(\lambda T)^j/j!][G^{(j)}(x) - G^{(j+1)}(x)]}{\sum_{j=0}^{\infty}[(\lambda T)^j/j!]G^{(j)}(x)}$$

is also strictly increasing in T for any $x > 0$ as follows: Differentiating $Q(T)$ with respect to T,

$$\frac{\lambda}{[\sum_{j=0}^{\infty}[(\lambda T)^j/j!]G^{(j)}(x)]^2}\left[\sum_{j=0}^{\infty}\frac{(\lambda T)^j}{j!}G^{(j+1)}(x)\sum_{i=0}^{\infty}\frac{(\lambda T)^i}{i!}G^{(i+1)}(x)\right.$$
$$\left. - \sum_{j=0}^{\infty}\frac{(\lambda T)^j}{j!}G^{(j)}(x)\sum_{i=0}^{\infty}\frac{(\lambda T)^i}{i!}G^{(i+2)}(x)\right].$$

The numerator is rewritten as

$$\sum_{j=0}^{\infty}\frac{(\lambda T)^j}{j!}\sum_{i=0}^{\infty}\frac{(\lambda T)^i}{i!}G^{(j)}(x)G^{(i+1)}(x)\left[\frac{G^{(j+1)}(x)}{G^{(j)}(x)} - \frac{G^{(i+2)}(x)}{G^{(i+1)}(x)}\right]$$
$$= \sum_{j=1}^{\infty}\frac{(\lambda T)^j}{j!}\sum_{i=0}^{j-1}\frac{(\lambda T)^i}{i!}G^{(j)}(x)G^{(i+1)}(x)\left[\frac{G^{(j+1)}(x)}{G^{(j)}(x)} - \frac{G^{(i+2)}(x)}{G^{(i+1)}(x)}\right]$$
$$+ \sum_{j=0}^{\infty}\frac{(\lambda T)^j}{j!}\sum_{i=j}^{\infty}\frac{(\lambda T)^i}{i!}G^{(j)}(x)G^{(i+1)}(x)\left[\frac{G^{(j+1)}(x)}{G^{(j)}(x)} - \frac{G^{(i+2)}(x)}{G^{(i+1)}(x)}\right]. \quad (3.35)$$

It can be easily seen that the second term on the right-hand side of (3.35) is positive because $G^{(j+1)}(x)/G^{(j)}(x)$ is strictly decreasing. Changing the summation of i and j, the first term on the right-hand side is

$$\sum_{i=0}^{\infty}\frac{(\lambda T)^i}{i!}\sum_{j=i+1}^{\infty}\frac{(\lambda T)^j}{j!}G^{(j)}(x)G^{(i+1)}(x)\left[\frac{G^{(j+1)}(x)}{G^{(j)}(x)} - \frac{G^{(i+2)}(x)}{G^{(i+1)}(x)}\right].$$

Changing i into j with each other, the above equation is

$$\sum_{j=0}^{\infty} \frac{(\lambda T)^j}{j!} \sum_{i=j+1}^{\infty} \frac{(\lambda T)^i}{i!} G^{(i)}(x) G^{(j+1)}(x) \left[\frac{G^{(i+1)}(x)}{G^{(i)}(x)} - \frac{G^{(j+2)}(x)}{G^{(j+1)}(x)} \right]$$

$$= \sum_{j=1}^{\infty} \frac{(\lambda T)^{j-1}}{(j-1)!} \sum_{i=j}^{\infty} \frac{(\lambda T)^{i+1}}{(i+1)!} G^{(i+1)}(x) G^{(j)}(x) \left[\frac{G^{(i+2)}(x)}{G^{(i+1)}(x)} - \frac{G^{(j+1)}(x)}{G^{(j)}(x)} \right].$$

Consequently, (3.35) is

$$\sum_{j=0}^{\infty} \frac{(\lambda T)^j}{j!} \sum_{i=j}^{\infty} \frac{(\lambda T)^i}{(i+1)!} G^{(j)}(x) G^{(i+1)}(x) \left[\frac{G^{(j+1)}(x)}{G^{(j)}(x)} - \frac{G^{(i+2)}(x)}{G^{(i+1)}(x)} \right] (i+1-j) > 0,$$

that completes the proof of that $Q(T)$ is strictly increasing.

From the above results, $Q(T)$ is strictly increasing from $e^{-\mu K}$ to 1 when $G(x) = 1 - e^{-\mu x}$. Thus, the left-hand side of (3.34) is also strictly increasing from 0 to μK. Therefore, if $c_K > c_T[1 + (1/\mu K)] + c_0/\mu$, then there exists a finite and unique T^* that satisfies (3.34), and the resulting cost rate is

$$\frac{C_1(T^*)}{\lambda} = \left(c_K - c_T - \frac{c_0}{\mu} \right) Q_1(T^*) + \frac{c_0}{\mu}. \tag{3.36}$$

Conversely, if $c_K \le c_T[1 + (1/\mu K)] + c_0/\mu$, then $T^* = \infty$.

The expected cost rate for the replacement at shock N is, from (3.32),

$$\frac{C_2(N)}{\lambda} \equiv \lim_{\substack{T \to \infty \\ Z \to \infty}} \frac{C(T, N, Z)}{\lambda}$$

$$= \frac{c_K - c_0/\mu - (c_K - c_N - c_0/\mu) G^{(N)}(K)}{\sum_{j=0}^{N-1} G^{(j)}(K)} + \frac{c_0}{\mu}$$

$$(N = 1, 2, \dots), \tag{3.37}$$

that agrees with (3.20) in the exponential case by replacing c_K with $c_K - c_0/\mu$. Because $Q_2(N)$ is strictly increasing to 1, if $c_K > c_N[1 + (1/\mu K)] + c_0\mu$, then there exists a finite and unique minimum N^* that minimizes $C_2(N)$.

Finally, the expected cost rate for the replacement at damage Z is, from (3.32),

$$\frac{C_3(Z)}{\lambda} \equiv \lim_{\substack{T \to \infty \\ N \to \infty}} \frac{C(T, N, Z)}{\lambda}$$

$$= \frac{c_K - c_0/\mu - (c_K - c_Z - c_0/\mu)(1 - e^{-\mu(K-Z)})}{1 + \mu Z} + \frac{c_0}{\mu}. \tag{3.38}$$

Differentiating $C_3(Z)$ with respect to Z and setting it equal to zero, for $c_K > c_Z + c_0/\mu$,

$$\mu Z e^{-\mu(K-Z)} = \frac{c_Z}{c_K - c_Z - c_0/\mu}. \tag{3.39}$$

The left-hand side of (3.39) is strictly increasing from 0 to μK. Thus, if $c_K > c_Z[1 + (1/\mu K)] + c_0/\mu$, then there exists a finite and unique Z^* ($0 < Z^* < K$) that satisfies (3.39), and the resulting cost rate is

$$\frac{C_3(Z^*)}{\lambda} = \frac{1}{\mu}\left(\frac{c_Z}{Z^*} + c_0\right). \tag{3.40}$$

It is of great interest that the condition that a finite optimum value exists is given by the same form as $c_K > c_k[1 + (1/\mu K)] + c_0/\mu$ ($k = T, N, Z$). In general, μK would be greater than $c_k/(c_K - c_k - c_0/\mu)$ because μK represents the expected number of shocks before failure.

Example 3.2. We compute the optimum T^*, N^*, and Z^* numerically. Table 3.2 presents the optimum λT^*, the expected cost rate $C_1(T^*)/(\lambda c_T)$, and $T^*/E\{Y\} = \lambda T^*/(1+\mu K)$ (see Example 2.2) for $\mu K = 10, 20$ and $c_K/c_T = 2$, 5, 10, 20 when $c_0 = 0$. If cost c_0 takes some positive value, then c_K may be replaced with $c_K - c_0/\mu$. Furthermore, the ratio of c_K to c_T becomes one indicator of replacement time. We compute $\widetilde{T}$ that satisfies $c_T/c_K = T/E\{Y\}$, i.e., $\lambda\widetilde{T} = (c_T/c_K)(1 + \mu K)$. This indicates that when $c_K/c_T = 2$, the unit should be replaced before failure at time $\lambda T^* = 9.02$ and 82.0% of the mean failure time. However, the approximate values $\widetilde{T}$ are too small to compare T^*, and hence, it would be useless practically.

Table 3.3 presents the optimum N^*, the expected cost rate $C_2(N^*)/(\lambda c_N)$, and $N^*/[1 + M_G(K)] = N^*/(1 + \mu K)$ for $\mu K = 10, 20$ and $c_N/c_T = 2, 5$, 10, 20. In addition, we compute a minimum $\widetilde{N}$ that satisfies $c_K \overline{G}^{(N)}(K) \geq c_N G^{(N)}(K)$. If the unit fails until the Nth shock, then it costs c_K, and otherwise, it costs c_N. The approximate values $\widetilde{N}$ show good upper bounds of N^* when $\mu K = 10$.

Table 3.4 presents the optimum μZ^*, the expected cost rate $C_3(Z^*)/(\lambda c_Z)$, and $Z^*/(K+1/\mu)$ for $\mu K = 10, 20$ and $c_Z/c_K = 2, 5, 10, 20$. Furthermore, we compute $\mu\widetilde{Z}$ that satisfies $c_Z/c_K = Z/(K+1/\mu)$, i.e., $\mu\widetilde{Z} = (c_Z/c_K)(1+\mu K)$ that agrees with $\lambda\widetilde{T}$ when $c_Z = c_T$, and $Z^* > \widetilde{Z}$. The expected costs $C_3(Z^*)$ are the smallest among three policies, as one expected. If costs c_K/c_k ($k = T$, N, Z) are the same ones, the replacement policy where the unit is replaced at damage Z is the best among the three policies. ∎

If the replacement cost is $c_K + c_0(x)$ ($x \geq K$) when the total damage is x and the unit is replaced at failure, then the expected cost rate in (3.32) is easily rewritten as

Table 3.2. Optimum time λT^*, expected cost rate $C_1(T^*)/(\lambda c_T)$, $T^*/E\{Y\}$, and approximate value $\lambda \widetilde{T}$ for c_K/c_T and μK

c_K/c_T	$\mu K = 10$			
	λT^*	$C_1(T^*)/(\lambda c_T)$	$\lambda T^*/(1+\mu K)$	$\lambda \widetilde{T}$
2	9.02	0.142	0.820	5.5
5	5.56	0.243	0.505	2.2
10	4.34	0.327	0.394	1.1
20	3.45	0.417	0.313	0.55
c_K/c_T	$\mu K = 20$			
	λT^*	$C_1(T^*)/(\lambda c_T)$	$\lambda T^*/(1+\mu K)$	$\lambda \widetilde{T}$
2	15.74	0.066	0.749	10.5
5	11.30	0.089	0.538	4.2
10	9.59	0.106	0.457	2.1
20	8.33	0.122	0.400	1.05

Table 3.3. Optimum number N^*, expected cost rate $C_2(N^*)/(\lambda c_N)$, $N^*/(1+\mu K)$, and approximate value $\widetilde{N}$ for c_K/c_N and μK

c_K/c_N	$\mu K = 10$			
	N^*	$C_2(N^*)/(\lambda c_N)$	$N^*/(1+\mu K)$	$\widetilde{N}$
2	9	0.156	0.818	10
5	6	0.213	0.545	8
10	5	0.253	0.455	7
20	4	0.300	0.364	6
c_K/c_N	$\mu K = 20$			
	N^*	$C_2(N^*)/(\lambda c_N)$	$N^*/(1+\mu K)$	$\widetilde{N}$
2	16	0.073	0.762	19
5	13	0.089	0.610	17
10	12	0.100	0.571	15
20	10	0.110	0.476	14

$$C(T,N,Z) = \frac{\begin{aligned} &c_K - (c_K - c_T)\sum_{j=0}^{N-1}[F_j(T) - F_{j+1}(T)]G^{(j)}(Z) \\ &- (c_K - c_N)F_N(T)G^{(N)}(Z) \\ &- (c_K - c_Z)\sum_{j=0}^{N-1}F_{j+1}(T)\int_0^Z[G(K-x) - G(Z-x)]\,dG^{(j)}(x) \\ &+ \sum_{j=0}^{N-1}F_{j+1}(T)\int_0^Z[\int_x^\infty \overline{G}(y-x)\,dc_0(y)]\,dG^{(j)}(x) \end{aligned}}{\sum_{j=0}^{N-1}G^{(j)}(Z)\int_0^T[F_j(t) - F_{j+1}(t)]\,dt}.$$

$$(3.41)$$

Table 3.4. Optimum damage level μZ^*, expected cost rate $C_3(Z^*)/(\lambda c_Z)$, $\mu Z^*/(1+\mu K)$, and approximate value $\widetilde{Z}$ for c_K/c_Z and μK

c_K/c_Z	μZ^*	$C_3(Z^*)/(\lambda c_Z)$	$\mu Z^*/(1+\mu K)$	$\mu \widetilde{Z}$
		$\mu K = 10$		
2	7.93	0.126	0.721	5.5
5	6.71	0.149	0.610	2.2
10	6.01	0.166	0.546	1.1
20	5.37	0.186	0.489	0.55
		$\mu K = 20$		
2	17.16	0.058	0.817	10.5
5	15.85	0.063	0.755	4.2
10	15.09	0.066	0.719	2.1
20	14.39	0.069	0.685	1.05

(2) Replacement at Time T or Damage Z

A unit is replaced before failure at time T or at damage Z, whichever occurs first. Then, the expected cost rate when $c_T = c_Z$ is, from (3.8),

$$C(T,Z) = \frac{c_T + (c_K - c_T)\sum_{j=0}^{\infty} F_{j+1}(T)\int_0^Z \overline{G}(K-x)\,dG^{(j)}(x)}{\sum_{j=0}^{\infty} G^{(j)}(Z)\int_0^T [F_j(t) - F_{j+1}(t)]\,dt}. \tag{3.42}$$

Let $f_j(t)$ and $g^{(j)}(x)$ be the density functions of $F_j(t)$ and $G^{(j)}(x)$, respectively. Differentiating $C(T,Z)$ with respect to T and setting it equal to zero,

$$Q_1(T,Z)\sum_{j=0}^{\infty} G^{(j)}(Z)\int_0^T [F_j(t) - F_{j+1}(t)]\,dt$$

$$-\sum_{j=0}^{\infty} F_{j+1}(T)\int_0^Z \overline{G}(K-x)\,dG^{(j)}(x) = \frac{c_T}{c_K - c_T}, \tag{3.43}$$

where

$$Q_1(T,Z) \equiv \frac{\sum_{j=0}^{\infty} f_{j+1}(T)\int_0^Z \overline{G}(K-x)\,dG^{(j)}(x)}{\sum_{j=0}^{\infty} G^{(j)}(Z)[F_j(T) - F_{j+1}(T)]}.$$

Furthermore, differentiating $C(T,Z)$ with respect to Z and setting it equal to zero,

$$Q_2(T,Z)\overline{G}(K-Z)\sum_{j=0}^{\infty}G^{(j)}(Z)\int_0^T[F_j(t)-F_{j+1}(t)]\,\mathrm{d}t$$

$$-\sum_{j=0}^{\infty}F_{j+1}(T)\int_0^Z\overline{G}(K-x)\,\mathrm{d}G^{(j)}(x)=\frac{c_T}{c_K-c_T},\qquad(3.44)$$

where

$$Q_2(T,Z)\equiv\frac{\sum_{j=1}^{\infty}g^{(j)}(Z)F_{j+1}(T)}{\sum_{j=1}^{\infty}g^{(j)}(Z)\int_0^T[F_j(t)-F_{j+1}(t)]\,\mathrm{d}t}.$$

In particular, when shocks occur in a Poisson process with rate λ, *i.e.*, $F_j(t)=\sum_{i=j}^{\infty}[(\lambda t)^i/i!]e^{-\lambda t}$, (3.43) and (3.44) are simplified, respectively, as follows:

$$Q_3(T,Z)\sum_{j=0}^{\infty}F_{j+1}(T)G^{(j)}(Z)$$

$$-\sum_{j=0}^{\infty}F_{j+1}(T)\int_0^Z\overline{G}(K-x)\,\mathrm{d}G^{(j)}(x)=\frac{c_T}{c_K-c_T},\qquad(3.45)$$

where

$$Q_3(T,Z)\equiv\frac{\sum_{j=0}^{\infty}[F_j(T)-F_{j+1}(T)]\int_0^Z\overline{G}(K-x)\,\mathrm{d}G^{(j)}(x)}{\sum_{j=0}^{\infty}[F_j(T)-F_{j+1}(T)]G^{(j)}(Z)},$$

and

$$\overline{G}(K-Z)\sum_{j=0}^{\infty}F_{j+1}(T)G^{(j)}(Z)$$

$$-\sum_{j=0}^{\infty}F_{j+1}(T)\int_0^Z\overline{G}(K-x)\,\mathrm{d}G^{(j)}(x)=\frac{c_T}{c_K-c_T}.\qquad(3.46)$$

Hence, there does not exist both T^* $(0<T^*<\infty)$ and Z^* $(0<Z^*<K)$ that satisfy (3.45) and (3.46) simultaneously, because $Q_3(T,Z)<\overline{G}(K-Z)$ for $T>0$, so that we may determine optimum T^* and Z^* independently under these conditions as shown in Section 3.2, and adopt the policy with a lower cost.

(3) Replacement at the Next Shock over Time T

It may be wasteful to replace an operating unit at planned times even if it is working. For example, when a unit is functioning for jobs with a variable working cycle and processing time, it would be better to do some maintenance after it has completed the work and process. The modified replacement model where a unit is replaced at the next failure after time T was considered [169], and the random maintenance model where it is replaced at random times was proposed in Section 9.3 of [1].

We consider the following modified replacement model: A unit is replaced before time T when the total damage has exceeded a failure level K, and after T, it is replaced at the next shock. Then, the probability that the unit is replaced before failure is

$$P_T = \sum_{j=0}^{\infty} [F_j(T) - F_{j+1}(T)] G^{(j+1)}(K), \qquad (3.47)$$

and the probability that it is replaced at failure is

$$P_K = \sum_{j=0}^{\infty} F_j(T)[G^{(j)}(K) - G^{(j+1)}(K)], \qquad (3.48)$$

where note that $(3.47) + (3.48) = 1$. The mean time to replacement is, from (3.47) and (3.48),

$$\sum_{j=0}^{\infty} G^{(j+1)}(K) \int_0^T \left[\int_{T-u}^{\infty} (t+u)\, \mathrm{d}F(t) \right] \mathrm{d}F_j(u)$$

$$+ \sum_{j=0}^{\infty} [G^{(j)}(K) - G^{(j+1)}(K)] \left\{ \int_0^T \left[\int_{T-u}^{\infty} (t+u)\, \mathrm{d}F(t) \right] \mathrm{d}F_j(u) + \int_0^T t\, \mathrm{d}F_{j+1}(t) \right\}$$

$$= \frac{1}{\lambda} \sum_{j=0}^{\infty} G^{(j)}(K) F_j(T). \qquad (3.49)$$

Therefore, the expected cost rate is

$$\frac{\tilde{C}_1(T)}{\lambda} = \frac{c_T P_T + c_K P_K}{\sum_{j=0}^{\infty} G^{(j)}(K) F_j(T)}$$

$$= \frac{c_K - (c_K - c_T) \sum_{j=0}^{\infty} [F_j(T) - F_{j+1}(T)] G^{(j+1)}(K)}{\sum_{j=0}^{\infty} F_j(T) G^{(j)}(K)}. \qquad (3.50)$$

When $T = 0$, $\tilde{C}_1(0)$ agrees with $C_2(1)$ in (3.21).

We derive an optimum time T^* that minimizes $\tilde{C}_1(T)$ when $F(t) = 1 - e^{-\lambda t}$ and $G(x) = 1 - e^{-\mu x}$, i.e., $F_j(t) = \sum_{i=j}^{\infty} [(\lambda t)^i / i!] e^{-\lambda t}$ and $G^{(j)}(x) = \sum_{i=j}^{\infty} [(\mu x)^i / i!] e^{-\mu x}$. Then, differentiating $\tilde{C}_1(T)$ in (3.50) with respect to T and setting it equal to zero,

$$\tilde{Q}(T) \sum_{j=0}^{\infty} F_j(T) G^{(j)}(K) - \sum_{j=0}^{\infty} \frac{(\lambda T)^j}{j!} e^{-\lambda T} [1 - G^{(j+1)}(K)] = \frac{c_T}{c_K - c_T}, \qquad (3.51)$$

where
$$\tilde{Q}(T) \equiv \frac{\sum_{j=0}^{\infty}[(\lambda T)^j/j!]e^{-\lambda T}[(\mu K)^{j+1}/(j+1)!]e^{-\mu K}}{\sum_{j=0}^{\infty}[(\lambda T)^j/j!]e^{-\lambda T}G^{(j+1)}(K)}.$$

Because $\tilde{Q}(T)$ is strictly increasing in T from $\mu K/(e^{\mu K}-1)$ to 1, the left-hand side of (3.51) is also strictly increasing from

$$D \equiv \frac{\mu K - 1 + e^{-\mu K}}{e^{\mu K} - 1} \le \frac{\mu K}{2}$$

to μK.

Therefore, we have the following optimum policy:

(i) If $D \ge c_T/(c_K - c_T)$, then $T^* = 0$, i.e., the unit is replaced at the first shock, and the expected cost rate is given in (3.21).

(ii) If $D < c_T/(c_K - c_T) < \mu K$, then there exists a finite and unique T^* that satisfies (3.51), and the resulting cost rate is

$$\tilde{C}_1(T^*) = \lambda(c_K - c_T)\tilde{Q}(T^*). \tag{3.52}$$

(iii) If $\mu K \le c_T/(c_K - c_T)$, then $T^* = \infty$, i.e., the unit is replaced only at failure, and the expected cost rate is given in (3.12).

(4) Replacement at the Next Shock over Damage Z

A unit is checked at each shock and the total damage is investigated only through inspection. If needed, it is replaced, as shown in (3) of Section 3.2. In addition, it may be better to replace a unit at the next shock time for prepare parts, workers, maintenance plans, and so on.

A unit is replaced when the total damage has exceeded a failure level K, and is also replaced at the next shock when the damage is between Z and K. Then, the probability that the unit is replaced between Z and K is

$$P_Z = \sum_{j=0}^{\infty} \int_0^Z \left[\int_{Z-x}^{K-x} G(K - x - y)\,dG(y) \right] dG^{(j)}(x), \tag{3.53}$$

and the probability that it is replaced when the total damage has exceeded K is

$$P_K = \sum_{j=0}^{\infty} \int_0^Z \left[\int_{Z-x}^{K-x} \overline{G}(K - x - y)\,dG(y) + \overline{G}(K - x) \right] dG^{(j)}(x), \tag{3.54}$$

where (3.53) + (3.54) = 1. Furthermore, the mean time to replacement is, from (3.53) and (3.54),

$$\frac{1}{\lambda} \sum_{j=0}^{\infty} \left\{ (j+2) \int_0^Z \left[\int_{Z-x}^{K-x} G(K-x-y)\, dG(y) \right] dG^{(j)}(x) \right.$$

$$\left. + \int_0^Z \left[(j+2) \int_{Z-x}^{K-x} \overline{G}(K-x-y)\, dG(y) + (j+1)\overline{G}(K-x) \right] dG^{(j)}(x) \right\}$$

$$= \frac{1}{\lambda} \left[1 + G(K) + \int_0^Z G(K-x)\, dM_G(x) \right]. \tag{3.55}$$

Therefore, the expected cost rate is

$$\frac{\tilde{C}_3(Z)}{\lambda} = \frac{c_Z P_Z + c_K P_K}{1 + G(K) + \int_0^Z G(K-x)\, dM_G(x)}$$

$$= \frac{c_K - (c_K - c_Z) \left\{ \int_Z^K G(K-x)\, dG(x) + \int_0^Z \left[\int_{Z-x}^{K-x} G(K-x-y)\, dG(y) \right] dM_G(x) \right\}}{1 + G(K) + \int_0^Z G(K-x)\, dM_G(x)}. \tag{3.56}$$

In particular, when $G(x) = 1 - e^{-\mu x}$,

$$\frac{\tilde{C}_3(Z)}{\lambda} = \frac{c_K - (c_K - c_Z)\{1 - [1 + \mu(K-Z)]e^{-\mu(K-Z)}\}}{1 + \mu Z + 1 - e^{-\mu(K-Z)}}. \tag{3.57}$$

Differentiating $\tilde{C}_3(Z)$ with respect to Z and setting it equal to zero,

$$e^{-\mu(K-Z)} \left[\frac{(1+\mu Z)\mu(K-Z)}{1 - e^{-\mu(K-Z)}} - 1 \right] = \frac{c_Z}{c_K - c_Z}. \tag{3.58}$$

The left-hand side of (3.58) is strictly increasing in Z from D to μK. Therefore, we have the following optimum policy:

(i) If $D \geq c_Z/(c_K - c_Z)$, then $Z^* = 0$, and the expected cost rate is

$$\frac{\tilde{C}_3(0)}{\lambda} = \frac{c_K - (c_K - c_Z)[1 - (1 + \mu K)e^{-\mu K}]}{2 - e^{-\mu K}}. \tag{3.59}$$

(ii) If $D < c_Z/(c_K - c_Z) < \mu K$, then there exists a finite and unique Z^* $(0 < Z^* < K)$ that satisfies (3.58), and the expected cost rate is

$$\frac{\tilde{C}_3(Z^*)}{\lambda} = \frac{(c_K - c_Z)\mu(K - Z^*)e^{-\mu(K-Z^*)}}{1 - e^{-\mu(K-Z^*)}}. \tag{3.60}$$

(iii) If $\mu K \leq c_Z/(c_K - c_Z)$, then $Z^* = K$, and the expected cost rate is given in (3.12).

It is of great interest that the condition for an optimum Z^* to exist is the same as that of **(3)**. Furthermore, compared (3.58) with (3.39), because

$$\frac{(1+\mu Z)\mu(K-Z)}{1-e^{-\mu(K-Z)}} > 1+\mu Z,$$

the optimum Z^* to satisfy (3.58) is smaller than that to satisfy (3.39), as one expected.

(5) Replacement at n Damage Levels

A unit is replaced before failure at damage Z_i $(i = 1, 2, \ldots, n)$, where $Z_{n+1} \equiv K$, and its replacement cost is c_i. Then, the probability that the unit is replaced at damage Z_i is

$$P_i = \sum_{j=0}^{\infty} \int_0^{Z_1} [G(Z_{i+1} - x) - G(Z_i - x)]\, dG^{(j)}(x)$$

$$= G(Z_{i+1}) - G(Z_i) + \int_0^{Z_1} [G(Z_{i+1} - x) - G(Z_i - x)]\, dM_G(x)$$

$$(i = 1, 2, \ldots, n), \qquad (3.61)$$

and the probability that it is replaced at failure is

$$P_K = \overline{G}(K) + \int_0^{Z_1} \overline{G}(K - x)\, dM_G(x), \qquad (3.62)$$

where note that $\sum_{i=1}^{n} P_i + P_K = 1$. Because the mean time to replacement is given by the denominator of (3.24), the expected cost rate is

$$\frac{C(Z_1, Z_2, \cdots, Z_n)}{\lambda} = \frac{c_K P_K + \sum_{i=1}^{n} c_i P_i}{1 + M_G(Z_1)}$$

$$= \frac{c_K - \sum_{i=1}^{n}(c_K - c_i)\left\{G(Z_{i+1}) - G(Z_i) + \int_0^{Z_1}[G(Z_{i+1} - x) - G(Z_i - x)]dM_G(x)\right\}}{1 + M_G(Z_1)},$$

$$(3.63)$$

that agrees with (3.24) for $n = 1$ when $Z_1 = Z$.

Next, a unit fails when the total damage has exceeded a failure level K_i, where $K_i < K_{i+1}$ and $K_\infty \equiv \infty$ $(i = 1, 2, \ldots)$, and its required cost is c_i with $c_i \le c_{i+1}$. If the unit is replaced before at damage Z $(Z \le K_1)$, then its probability is

$$P_Z = \sum_{j=0}^{\infty} \int_0^{Z} [G(K_1 - x) - G(Z - x)]\, dG^{(j)}(x), \qquad (3.64)$$

and the probability that it is replaced at failure level K_i is

$$P_i = \sum_{j=0}^{\infty} \int_0^Z [G(K_{i+1} - x) - G(K_i - x)] \, dG^{(j)}(x) \quad (i = 1, 2, \dots), \quad (3.65)$$

where $P_Z + \sum_{i=1}^{\infty} P_i = 1$. Thus, the expected cost rate is

$$\begin{aligned} \frac{C(Z)}{\lambda} &= \frac{c_Z P_Z + \sum_{i=1}^{\infty} c_i P_i}{1 + M_G(Z)} \\ &= \frac{c_Z + \sum_{i=1}^{\infty}(c_i - c_Z)\{G(K_{i+1}) - G(K_i)}{1 + M_G(Z)} \\ &\quad \frac{+ \int_0^Z [G(K_{i+1} - x) - G(K_i - x)] \, dM_G(x)\}}{1 + M_G(Z)}. \end{aligned} \quad (3.66)$$

Differentiating $C(Z)$ with respect to Z and setting it equal to zero,

$$\sum_{i=1}^{\infty} (c_i - c_{i-1}) \int_{K_i - Z}^{K_i} [1 + M_G(K_i - x)] \, dG(x) = c_Z, \quad (3.67)$$

where $c_0 \equiv c_Z < c_1$. Thus, if $M_G(K_1) > c_Z/(c_1 - c_Z)$, then there exists a finite and unique Z^* ($0 < Z^* < K_1$) that satisfies (3.67), and it is smaller than that to satisfy (3.25).

(6) Random Replacement Interval

Suppose that a unit is also replaced at random time R with a general distribution $\gamma(t)$ for the same policy in Section 3.1. This corresponds to the model where a unit is replaced at the same random times as its working times (see Section 9.3 in [1]).

By a method similar to obtaining (3.1)–(3.4), the probability that the unit is replaced at time T is

$$P_T = \sum_{j=0}^{N-1} \overline{\gamma}(T)[F_j(T) - F_{j+1}(T)]G^{(j)}(Z), \quad (3.68)$$

the probability that it is replaced at shock N is

$$P_N = \int_0^T \overline{\gamma}(t) \, dF_N(t) G^{(N)}(Z), \quad (3.69)$$

the probability that it is replaced at damage Z is

$$P_Z = \sum_{j=0}^{N-1} \int_0^T \overline{\gamma}(t) \, dF_{j+1}(t) \int_0^Z [G(K - x) - G(Z - x)] \, dG^{(j)}(x), \quad (3.70)$$

the probability that it is replaced at damage K is

$$P_K = \sum_{j=0}^{N-1} \int_0^T \overline{\gamma}(t) \, dF_{j+1}(t) \int_0^Z \overline{G}(K-x) \, dG^{(j)}(x), \qquad (3.71)$$

and the probability that it is replaced at random time R is

$$P_R = \sum_{j=0}^{N-1} \int_0^T [F_j(t) - F_{j+1}(t)] \, d\gamma(t) G^{(j)}(Z), \qquad (3.72)$$

where $\overline{\gamma}(t) \equiv 1 - \gamma(t)$ and $P_T + P_N + P_Z + P_K + P_R = 1$. Similarly, the mean time to replacement is

$$T \sum_{j=0}^{N-1} \overline{\gamma}(T)[F_j(T) - F_{j+1}(T)]G^{(j)}(Z) + \int_0^T t\,\overline{\gamma}(t) \, dF_N(t)G^{(N)}(Z)$$

$$+ \sum_{j=0}^{N-1} \int_0^T t\,\overline{\gamma}(t) \, dF_{j+1}(t) \int_0^Z [G(K-x) - G(Z-x)] \, dG^{(j)}(x)$$

$$+ \sum_{j=0}^{N-1} \int_0^T t\,\overline{\gamma}(t) \, dF_{j+1}(t) \int_0^Z \overline{G}(K-x) \, dG^{(j)}(x)$$

$$+ \sum_{j=0}^{N-1} \int_0^T t\,[F_j(t) - F_{j+1}(t)] \, d\gamma(t)G^{(j)}(Z)$$

$$= \sum_{j=0}^{N-1} G^{(j)}(Z) \int_0^T \overline{\gamma}(t)[F_j(t) - F_{j+1}(t)] \, dt. \qquad (3.73)$$

Let c_R be the replacement cost at random time R and c_T, c_N, c_Z, and c_K be the same costs given in (3.6). Then, the expected cost rate is

$$C(T, N, Z, R) = \frac{c_T P_T + c_N P_N + c_Z P_Z + c_K P_K + c_R P_R}{\sum_{j=0}^{N-1} G^{(j)}(Z) \int_0^T \overline{\gamma}(t)[F_j(t) - F_{j+1}(t)] \, dt}, \qquad (3.74)$$

that agrees with (3.8) when $\overline{\gamma}(t) \equiv 1$.

4

Replacement of Multiunit Systems

In general, a system consists of a variety of units. In (4) of Section 2.4, we have considered a system with n different units and derived the first-passage time distributions to system failure. If a system consists of a series system, then we may consider a maintenance policy before the first failure of units. If a system consists of a parallel system, then we may consider a maintenance policy before the last failure of units. But, in general, it would be difficult to discuss analytically optimum maintenance policies for shock and damage models of multiunit systems. A conditioned-based maintenance of a two-unit series system whose deterioration is monitored at periodic times was considered, and its optimum policy was discussed, using dynamic programming [170].

In Section 4.1, we take up a parallel system with n identical units that are situated in a random environment, as shown in Example 1.2. Each unit fails successively from shocks in a random environment, and finally, the system fails when all units have failed at some shock. For such units, we consider the two cases where the probability of unit failure is constant at any shock and its probability depends on the number of shocks. As the preventive replacement, the system is replaced before system failure when the total number of failed units is $N+1, N+2, \cdots, n-1$ at some shock. Introducing replacement costs, we obtain the expected cost rates for the two cases and derive optimum numbers N^* that minimize them. Furthermore, we apply the replacement model to a damage model where each unit fails when the damage due to shocks has exceeded a failure level K. On the other hand, we consider the replacement model of a k-out-of-n system that is replaced at a shock number N and obtain the expected cost rate.

In multiunit redundant systems, the failure of some units may affect one or more of the remaining units. This is called *failure interaction*. Two types of induced failure and shock damage are defined [171]. In Section 4.2, we consider a two-unit system with unit 1 and unit 2, where unit 2 fails with some probability at the jth time of unit 1 failure (induced failure), and it causes an amount of damage to unit 2 (shock damage). As the replacement policy, the system is replaced at the Nth failure of unit 1 or at the failure of unit 2, whichever occurs first. We obtain the expected cost rates for the two types of failure interaction and derive optimum numbers N^* that minimize them. Furthermore, we propose two extended models where the system is replaced at a planned time T or (1) at the Nth failure of unit 1 and (2) at a damage level Z of unit 2.

4.1 Parallel System in a Random Environment

Consider a standard parallel redundant system that consists of n identical units and fails when all units have failed. The system is situated in a random environment that generates shocks according to a general distribution $F(t)$ with finite mean $1/\lambda$. Each unit fails from shocks, independently of the other units. The failure distribution and the mean time to system failure have been derived in Example 1.2.

We consider the following three cases: The probability that each unit fails is constant p at all shocks, the probability that it fails at the jth shock is $p(j)$ that depends on the number of shocks, and the probability that it fails until the jth shock is $1 - G^{(j)}(K)$. Then, the system is replaced before system failure when the total number of failed units is $N + 1, N + 2, \cdots, n - 1$, and it is replaced when all units have failed, otherwise, it is left alone. For such replacement models, we introduce the replacement costs: Cost c_n is incurred when the failed system is replaced, and cost c_N ($c_N < c_n$) is incurred when the system with m ($m = N + 1, N + 2, \cdots, n - 1$) failed units is replaced before system failure. Furthermore, we consider an additional replacement cost that is a linear function of failed units. Under these assumptions, we derive optimum numbers N^* that minimize the expected cost rates for the three models.

4.1.1 Replacement Model

Consider a parallel system with n (≥ 2) identical units, each of which fails at shocks with probability p ($0 < p \leq 1$), where $q \equiv 1 - p$ [52]. Shocks occur in a renewal process with mean interval time $1/\lambda$. Let W_j be the total number of units that fail at the jth ($j = 1, 2, \cdots$) shock, where $W_0 \equiv 0$. Then, the probability that the system is replaced after failure is

$$P_n \equiv \Pr\{W_1 = n\} + \sum_{j=2}^{\infty} \sum_{r=0}^{N} \Pr\{W_1 + W_2 + \cdots + W_{j-1} = r$$

$$\text{and } W_1 + W_2 + \cdots + W_j = n\}$$

$$= p^n + p^n \sum_{j=2}^{\infty} \sum_{r=0}^{N} \sum_{i_1+i_2+\cdots+i_{j-1}=r} \binom{n}{i_1} q^{n-i_1} \binom{n-i_1}{i_2} q^{n-i_1-i_2}$$

$$\cdots \binom{n-i_1-i_2-\cdots-i_{j-2}}{i_{j-1}} q^{n-i_1-i_2-\cdots-i_{j-1}}$$

$$= p^n + p^n \sum_{j=2}^{\infty} \sum_{r=0}^{N} \binom{n}{r} (q^j)^{n-r} (1 + q + \cdots + q^{j-1})^r$$

$$= \sum_{r=0}^{N} \binom{n}{r} p^{n-r} \sum_{i=0}^{r} \binom{r}{i} (-1)^i \sum_{j=0}^{\infty} (q^{n-r+i})^j$$

$$= \sum_{r=0}^{N} \binom{n}{r} (-1)^r p^{n-r} \sum_{i=0}^{r} \binom{r}{i} (-1)^i \frac{1}{1 - q^{n-i}}. \tag{4.1}$$

Similarly, the probability that the system is replaced before failure is

$$P_N \equiv \Pr\{N + 1 \le W_1 \le n - 1\}$$

$$+ \sum_{j=2}^{\infty} \sum_{r=0}^{N} \Pr\{W_1 + W_2 + \cdots + W_{j-1} = r$$

$$\text{and } N + 1 \le W_1 + W_2 + \cdots + W_j \le n - 1\}$$

$$= \sum_{r=N+1}^{n-1} \binom{n}{r} (-1)^r p^{n-r} \sum_{i=0}^{r} \binom{r}{i} (-1)^i \frac{1}{1 - q^{n-i}}, \tag{4.2}$$

where $P_n + P_N = 1$. For the derivations of (4.1) and (4.2), refer to the next sections.

Furthermore, the mean time to replacement, *i.e.*, the mean time that the total number of failed units has exceeded $N + 1$ for the first time at some shock is

$$l_{N+1} = \sum_{j=1}^{\infty} \frac{j}{\lambda} \sum_{r=0}^{N} \Pr\{W_1 + W_2 + \cdots + W_{j-1} = r$$

$$\text{and } W_1 + W_2 + \cdots + W_j \ge N + 1\}$$

$$= \frac{1}{\lambda} \sum_{r=0}^{N} \binom{n}{r} (-1)^r \sum_{i=0}^{N-r} \binom{n-r}{i} \frac{1}{1 - q^{n-i}} \quad (N = 0, 1, 2, \ldots, n - 1).$$

$$\tag{4.3}$$

It is also equal to the mean time to failure of an $(N+1)$-out-of-n system that fails if and only if at least $N+1$ of n units fail. In particular, when $N = n-1$,

(4.3) is simplified as

$$l_n = \frac{1}{\lambda} \sum_{i=1}^{n} \binom{n}{i} \frac{(-1)^{i+1}}{1 - q^i}, \tag{4.4}$$

that is the mean time to failure of an n-unit parallel system in Example 1.2.
Therefore, the expected cost rate is

$$\frac{C_1(N)}{\lambda} = \frac{c_n P_n + c_N P_N}{\lambda l_{N+1}}$$

$$= \frac{c_N + (c_n - c_N) \sum_{r=0}^{N} \binom{n}{r}(-1)^r p^{n-r} \sum_{i=0}^{r} \binom{r}{i}(-1)^i [1/(1 - q^{n-i})]}{\sum_{r=0}^{N} \binom{n}{r}(-1)^r \sum_{i=0}^{N-r} \binom{n-r}{i}[1/(1 - q^{n-i})]}$$

$$(N = 0, 1, 2, \cdots, n - 1). \tag{4.5}$$

It is evident that

$$\frac{C_1(n-1)}{\lambda} = \frac{c_n}{\sum_{i=1}^{n} \binom{n}{i}(-1)^{i+1}[1/(1 - q^i)]}, \tag{4.6}$$

$$\frac{C_1(0)}{\lambda} = c_n p^n + c_N(1 - p^n - q^n). \tag{4.7}$$

Thus, when the number n of units is given, we can determine an optimum number N^* that minimizes $C_1(N)$ by comparing it for $N = 0, 1, \cdots, n - 1$. For example, when $n = 2$,

$$\frac{C_1(0)}{\lambda} = c_n p^2 + 2c_N pq,$$

$$\frac{C_1(1)}{\lambda} = \frac{c_n(1 - q^2)}{1 + 2q}.$$

Hence, if $q/(1 + 2q) > c_N/c_n$, then $N^* = 0$, i.e., the system is replaced when only one unit has failed. If $q/(1+2q) \leq c_N/c_n$, then $N^* = 1$, i.e., it is replaced when two units have failed. In addition, because $q/(1+2q) \leq 1/3$, if $c_n \leq 3c_N$, then $N^* = 1$.

Example 4.1.   Table 4.1 presents the optimum number N^* for $n = 2, 4, 8,$ 15, 20 and $p = 0.01, 0.05, 0.10, 0.20, 0.30, 0.40, 0.50$ when $c_N/c_n = 0.1$. It is natural that the optimum N^* is decreasing in p and increasing in n. For example, if the total number of failed units is 6 or 7 at some shock when $n = 8$ and $p = 0.10$, then the system should be replaced before failure. In particular, when $n = 2$, if $p < 0.875$, then $N^* = 0$. ∎

4.1.2 Extended Replacement Models

It is assumed in the same model, as that of Section 4.1.1, that the probability that an operating unit fails at the jth shock is $p(j)$ $(j = 1, 2, \cdots)$, depending

Table 4.1. Optimum number N^* of a parallel system with n units when $c_N/c_n = 0.1$

p	n				
	2	4	8	15	20
0.01	0	2	6	13	18
0.05	0	2	6	13	18
0.10	0	2	5	12	17
0.20	0	2	5	12	17
0.30	0	1	5	11	16
0.40	0	1	4	11	16
0.50	0	1	4	10	15

on the number of shocks [53]. This assumption is more reasonable because the damage due to shocks would be additive and the failure rate would increase with time. In addition, cost $nc_0 + c_n$ is incurred when a failed system is replaced, where costs c_0 and c_n include all costs resulting from the failure and replacement of one unit and the system, respectively. Cost $mc_0 + c_N$ is incurred when m $(m = N+1, N+2, \cdots, n-1)$ units have failed and the system is replaced before its failure. Let $P(j) \equiv \sum_{i=1}^{j} p(i)$ $(j = 1, 2, \cdots)$ be the probability that each unit fails until the jth shock, where $P(0) \equiv 0$. First, by a method similar to obtaining (4.3), the mean time to system failure is

$$l_n = \sum_{j=1}^{\infty} \frac{j}{\lambda} \sum_{r=0}^{n-1} \Pr\{W_1 + W_2 + \cdots + W_{j-1} = r$$

$$\text{and } W_1 + W_2 + \cdots + W_j = n\}$$

$$= \sum_{j=1}^{\infty} \frac{j}{\lambda} \sum_{r=0}^{n-1} \binom{n}{r} [p(j)]^{n-r} [P(j-1)]^r$$

$$= \frac{1}{\lambda} \sum_{j=0}^{\infty} \{1 - [P(j)]^n\}$$

$$= \frac{1}{\lambda} \sum_{i=1}^{n} \binom{n}{i} (-1)^{i+1} \sum_{j=0}^{\infty} [\overline{P}(j)]^i, \tag{4.8}$$

where $\overline{P}(j) \equiv 1 - P(j)$. For example, when $\overline{P}(j) = (q)^{j^\alpha}$ $(\alpha > 0)$, i.e., each unit fails according to a discrete Weibull distribution (see Section 1.2 of [1]),

$$l_n = \frac{1}{\lambda} \sum_{i=1}^{n} \binom{n}{i} (-1)^{i+1} \sum_{j=0}^{\infty} [(q)^{j^\alpha}]^i. \tag{4.9}$$

In the particular case of $\alpha = 1$, l_n is equal to (4.4).

We obtain the expected cost rate. Let P_m be the probability that the total number of units failed at some shock becomes m $(m = N+1, N+2, \cdots, n)$

and hence, the system is replaced. Then,

$$P_m = \sum_{j=1}^{\infty} \sum_{r=0}^{N} \Pr\{W_1 + W_2 + \cdots + W_{j-1} = r$$

$$\text{and } W_1 + W_2 + \cdots + W_j = m\}$$

$$= \sum_{j=1}^{\infty} \binom{n}{m} [\overline{P}(j)]^{n-m} \sum_{r=0}^{N} \binom{m}{r} [p(j)]^{m-r} [P(j-1)]^r$$

$$(m = N+1, N+2, \ldots, n), \qquad (4.10)$$

where $\sum_{m=N+1}^{n} P_m = 1$. Furthermore, in a similar way of obtaining (4.8), the mean time to replacement, *i.e.*, the mean time that the total number of failed units exceeds $N+1$ for the first time is

$$l_{N+1} = \sum_{j=1}^{\infty} \frac{j}{\lambda} \sum_{m=N+1}^{n} \sum_{r=0}^{N} \Pr\{W_1 + W_2 + \cdots + W_{j-1} = r$$

$$\text{and } W_1 + W_2 + \cdots + W_j = m\}$$

$$= \sum_{j=1}^{\infty} \frac{j}{\lambda} \sum_{m=N+1}^{n} \binom{n}{m} [\overline{P}(j)]^{n-m} \sum_{r=0}^{N} \binom{m}{r} [p(j)]^{m-r} [P(j-1)]^r$$

$$= \frac{1}{\lambda} \sum_{j=0}^{\infty} \sum_{m=0}^{N} \binom{n}{m} [\overline{P}(j)]^{n-m} [P(j)]^m. \qquad (4.11)$$

Thus, the expected cost rate is

$$C_2(N) = \frac{(c_0 n + c_n) P_n + \sum_{m=N+1}^{n-1} (c_0 m + c_N) P_m}{\text{mean time to replacement}}$$

$$= \frac{c_N + (c_n - c_N) P_n + c_0 \sum_{m=N+1}^{n} m P_m}{l_{N+1}}.$$

Therefore, from (4.10) and (4.11),

$$\frac{C_2(N)}{\lambda} = \frac{\begin{array}{c} c_N + (c_n - c_N) \sum_{j=1}^{\infty} \sum_{r=0}^{N} \binom{n}{r} [p(j)]^{n-r} [P(j-1)]^r \\ + c_0 n \sum_{j=1}^{\infty} \sum_{m=N+1}^{n} \binom{n-1}{m-1} [\overline{P}(j)]^{n-m} \sum_{r=0}^{N} \binom{m}{r} [p(j)]^{m-r} [P(j-1)]^r \end{array}}{\sum_{j=0}^{\infty} \sum_{m=0}^{N} \binom{n}{m} [\overline{P}(j)]^{n-m} [P(j)]^m}$$

$$(N = 0, 1, 2, \cdots, n-1). \qquad (4.12)$$

It is clearly seen that

$$\frac{C_2(n-1)}{\lambda} = \frac{c_0 n + c_n}{\sum_{j=0}^{\infty} \{1 - [P(j)]^n\}}, \qquad (4.13)$$

$$\frac{C_2(0)}{\lambda} = \frac{c_N + (c_n - c_N) \sum_{j=1}^{\infty} [p(j)]^n + c_0 n \sum_{j=1}^{\infty} p(j) [\overline{P}(j-1)]^{n-1}}{\sum_{j=0}^{\infty} [\overline{P}(j)]^n},$$

$$(4.14)$$

Table 4.2. Optimum number N^* of a parallel system with n units when $c_0/c_n = 0.05$ and $c_N/c_n = 0.1$

p	n				
	2	4	8	15	20
0.01	0	2	7	14	19
0.05	0	2	6	14	19
0.10	0	2	6	14	19
0.20	0	2	6	13	19
0.30	0	1	5	13	18
0.40	0	1	5	12	18
0.50	0	1	5	12	17

that represents the expected cost for an n-unit parallel system and an n-unit series system when $c_n = c_N$, respectively.

If n and $p(j)$ are given, we can determine an optimum number N^* that minimizes the expected cost $C_2(N)$ in (4.12) by comparing $N = 0, 1, 2, \cdots, n-1$. If $p(j)$ is a geometric distribution, i.e., $p(j) = pq^{j-1}$ and $P(j) = 1 - q^j$ ($p \equiv 1 - q > 0$), then

$$
\frac{C_2(N)}{\lambda} = \frac{
\begin{aligned}
&c_N + (c_n - c_N) \sum_{r=0}^{N} \binom{n}{r}(-1)^r p^{n-r} \sum_{i=0}^{r} \binom{r}{i}(-1)^i [1/(1 - q^{n-i})] \\
&+ c_0 n p \sum_{r=0}^{N} \binom{n-1}{r}(-1)^r \sum_{i=0}^{N-r} \binom{n-1-r}{i}[1/(1 - q^{n-i})]
\end{aligned}
}{
\sum_{r=0}^{N} \binom{n}{r}(-1)^r \sum_{i=0}^{N-r} \binom{n-r}{i}[1/(1 - q^{n-i})]
}
$$
$$(N = 0, 1, 2, \cdots n - 1). \tag{4.15}$$

In this case, if $c_0 = 0$, then the above result agrees with (4.5).

Example 4.2.   Suppose that the failure distribution is a negative binomial distribution with a shape parameter 2, i.e., $p(j) = jp^2 q^{j-1}$ ($j = 1, 2, \cdots$) where $q \equiv 1 - p$. Table 4.2 presents the optimum number N^* that minimizes the expected cost $C_2(N)$ for several n and p when $c_0/c_n = 0.05$ and $c_N/c_n = 0.1$. This indicates that the values of N^* are not less than those of Table 4.1 for the same p and n. ∎

Next, we apply the previous replacement model to a damage model. Suppose that the total damage is not additive and each unit fails when the damage due to some shock has exceeded a failure level K. We consider an independent damage model discussed in Section 2.2: Shocks occur in a renewal process with finite mean $1/\lambda$. The damage W_j due to each shock has an identical distribution $G(x) \equiv \Pr\{W_j \le x\}$ and the total damage is not additive, i.e., each unit fails with probability $[G(K)]^{j-1} - [G(K)]^j$ at shock j ($j = 1, 2, \cdots$). Then, replacing $p = \overline{G}(K)$ formally in (4.5), the expected cost rate for a parallel system is

$$\frac{C_1(N)}{\lambda} = \frac{\begin{array}{c} c_N + (c_n - c_N) \sum_{r=0}^{N} \binom{n}{r}(-1)^r [\overline{G}(K)]^{n-r} \\ \times \sum_{i=0}^{r} \binom{r}{i}(-1)^i [1/\{1 - [G(K)]^{n-i}\}] \end{array}}{\sum_{r=0}^{N} \binom{n}{r}(-1)^r \sum_{i=0}^{N-r} \binom{n-r}{i}[1/\{1 - [G(K)]^{n-i}\}]}$$

$$(N = 0, 1, 2, \cdots, n-1). \qquad (4.16)$$

On the other hand, the total damage is additive, *i.e.*, each unit fails with probability $G^{(j-1)}(K) - G^{(j)}(K)$ at shock j $(j = 1, 2, \cdots)$. Then, replacing $p(j) = G^{(j-1)}(K) - G^{(j)}(K)$ and $P(j) = 1 - G^{(j)}(K)$ formally in (4.12), the expected cost rate is

$$\frac{C_2(N)}{\lambda} = \frac{\begin{array}{c} c_N + (c_n - c_N) \sum_{j=1}^{\infty} \sum_{r=0}^{N} \binom{n}{r}[G^{(j-1)}(K) - G^{(j)}(K)]^{n-r} \\ \times [1 - G^{(j-1)}(K)]^r + c_0 n \sum_{j=1}^{\infty} \sum_{m=N+1}^{\infty} \binom{n-1}{m-1}[G^{(j)}(K)]^{n-m} \\ \times \sum_{r=0}^{N} \binom{m}{r}[G^{(j-1)}(K) - G^{(j)}(K)]^{m-r}[1 - G^{(j-1)}(K)]^r \end{array}}{\sum_{j=0}^{\infty} \sum_{r=0}^{N} \binom{n}{r}[G^{(j)}(K)]^{n-r}[1 - G^{(j)}(K)]^r}$$

$$(N = 0, 1, 2, \cdots, n-1). \qquad (4.17)$$

4.1.3 Replacement at Shock Number

Suppose in the same model as that of Section 4.1.2 that the system is replaced at a shock number N $(N = 1, 2, \cdots)$ or at a system failure, whichever occurs first. Then, the probability that the system is replaced at failure until shock N is

$$\sum_{j=1}^{N} \sum_{r=0}^{n-1} \Pr\{W_1 + W_2 + \cdots + W_{j-1} = r \text{ and } W_1 + W_2 + \cdots + W_j = n\}$$

$$= \sum_{j=1}^{N} \sum_{r=0}^{n-1} \binom{n}{r}[p(j)]^{n-r}[P(j-1)]^r = [P(N)]^n, \qquad (4.18)$$

and the probability that it is replaced before failure at shock N is

$$\sum_{r=0}^{n-1} \Pr\{W_1 + W_2 + \cdots + W_N = r\} = \sum_{r=0}^{n-1} \binom{n}{r}[\overline{P}(N)]^{n-r}[P(N)]^r$$

$$= 1 - [P(N)]^n. \qquad (4.19)$$

Similarly, the mean time to replacement is

$$\sum_{j=1}^{N} \frac{j}{\lambda} \sum_{r=0}^{n-1} \binom{n}{r}[p(j)]^{n-r}[P(j-1)]^r + N\{1 - [P(N)]^n\}$$

$$= \frac{1}{\lambda} \sum_{j=0}^{N-1} \{1 - [P(j)]^n\}, \qquad (4.20)$$

and the expected number of failed units until replacement is

$$\sum_{r=0}^{n} r \binom{n}{r} [\overline{P}(N)]^{n-r} [P(N)]^{r} = nP(N).$$

Therefore, the expected cost rate is

$$\frac{\widetilde{C}_2(N)}{\lambda} = \frac{c_N + (c_n - c_N)[P(N)]^n + c_0 n P(N)}{\sum_{j=0}^{N-1} \{1 - [P(j)]^n\}} \qquad (N = 1, 2, \cdots). \qquad (4.21)$$

Next, consider a k-out-of-n system that fails when the total number of failed units is more than k at some shock. Then, in a way similar to obtaining (4.21), the probability that the system is replaced at failure is

$$\sum_{j=1}^{N} \sum_{m=k+1}^{n} \sum_{r=0}^{k} \Pr\{W_1 + W_2 + \cdots + W_{j-1} = r$$

$$\text{and } W_1 + W_2 + \cdots + W_j = m\}$$

$$= \sum_{j=1}^{N} \sum_{m=k+1}^{n} \binom{n}{m} [\overline{P}(j)]^{n-m} \sum_{r=0}^{k} \binom{m}{r} [P(j-1)]^r [p(j)]^{m-r}$$

$$= \sum_{j=1}^{N} \sum_{m=k+1}^{n} \binom{n}{m} \{[\overline{P}(j)]^{n-m}[P(j)]^m - [\overline{P}(j-1)]^{n-m}[P(j-1)]^m\}$$

$$= \sum_{m=k+1}^{n} \binom{n}{m} [\overline{P}(N)]^{n-m}[P(N)]^m, \qquad (4.22)$$

and the probability that it is replaced at shock N is

$$\sum_{m=0}^{k} \sum_{r=0}^{k} \Pr\{W_1 + W_2 + \cdots + W_{N-1} = r \text{ and } W_1 + W_2 + \cdots + W_N = m\}$$

$$= \sum_{m=0}^{k} \binom{n}{m} [\overline{P}(N)]^{n-m}[P(N)]^m, \qquad (4.23)$$

so that the mean time to replacement is

$$\sum_{j=1}^{N} \frac{j}{\lambda} \sum_{m=k+1}^{n} \binom{n}{m} \{[\overline{P}(j)]^{n-m}[P(j)]^m - [\overline{P}(j-1)]^{n-m}[P(j-1)]^m\}$$

$$+ \frac{N}{\lambda} \sum_{m=0}^{k} \binom{n}{m} [\overline{P}(N)]^{n-m}[P(N)]^m$$

$$= \frac{1}{\lambda} \sum_{j=0}^{N-1} \sum_{m=0}^{k} \binom{n}{m} [\overline{P}(j)]^{n-m}[P(j)]^m, \qquad (4.24)$$

and the expected number of failed units until replacement is

$$\sum_{j=1}^{N} \sum_{m=k+1}^{n} m \sum_{r=0}^{k} \Pr\{W_1 + W_2 + \cdots + W_{j-1} = r$$

$$\text{and } W_1 + W_2 + \cdots + W_j = m\}$$

$$+ \sum_{m=0}^{k} m \sum_{r=0}^{k} \Pr\{W_1 + W_2 + \cdots + W_{N-1} = r$$

$$\text{and } W_1 + W_2 + \cdots + W_N = m\}$$

$$= \sum_{j=1}^{N} \left[\sum_{m=k+1}^{n} m \binom{n}{m} \left\{ [\overline{P}(j)]^{n-m}[P(j)]^m - [\overline{P}(j-1)]^{n-m}[P(j-1)]^m \right\} \right.$$

$$\left. - np(j) \sum_{m=k+1}^{n-1} \binom{n-1}{m} [\overline{P}(j-1)]^{n-1-m}[P(j-1)]^m \right]$$

$$+ \sum_{m=0}^{k} m \binom{n}{m} [\overline{P}(N)]^{n-m}[P(N)]^m$$

$$= n \sum_{j=0}^{N-1} p(j+1) \sum_{m=0}^{k} \binom{n-1}{m} [\overline{P}(j)]^{n-1-m}[P(j)]^m. \tag{4.25}$$

Therefore, the expected cost rate is, from (4.22), (4.24), and (4.25),

$$\frac{\widetilde{C}_2(N|k)}{\lambda} = \frac{c_N + (c_n - c_N) \sum_{m=k+1}^{n} \binom{n}{m} [\overline{P}(N)]^{n-m}[P(N)]^m}{\sum_{j=0}^{N-1} \sum_{m=0}^{k} \binom{n}{m} [\overline{P}(j)]^{n-m}[P(j)]^m}$$

$$(N = 1, 2, \dots). \tag{4.26}$$

In particular, when $k = n-1, \widetilde{C}_2(N|n-1)$ is equal to (4.21). Furthermore, when $k = 0$, i.e., the system consists of a series system, the expected cost rate is

$$\frac{\widetilde{C}_2(N|0)}{\lambda} = \frac{c_n - (c_n - c_N)[\overline{P}(N)]^n + c_0 n \sum_{j=0}^{N-1} p(j+1)[\overline{P}(j)]^{n-1}}{\sum_{j=0}^{N-1} [\overline{P}(j)]^n}. \tag{4.27}$$

Some modified replacement models for k-out-of-n systems [172–174] and consecutive k-out-of-n systems [175, 176] subject to shocks were proposed.

4.2 Two-unit System with Failure Interactions

In a multiunit system, the failure times of different units may be often statistically correlated [177]. In other instances, the failure of units can affect one or more of the remaining units. Such types of interactions between units have been termed *failure interaction* [171]. Two types of failure interactions such as induced failure and shock damage were defined, and the preventive maintenance of a two-unit system with shock damage interaction was considered [178].

This section considers a system with unit 1 and unit 2. If unit 1 fails then it undergoes only minimal repair, and hence, unit 1 failures occur in a non-homogeneous Poisson process with a mean value function $H(t) \equiv \int_0^t h(u)du$, where an intensity function $h(t)$ is increasing in t (see Section 4.1 of [1]).

Further, when unit 1 fails, we indicate the following two failure interactions between two units [54]:

(1) Induced failure: Unit 2 fails with probability α_j at the jth time of unit 1 failure.
(2) Shock damage: Unit 1 failure causes an amount of damage with distribution $G(x)$ to unit 2.

Suppose that the system is replaced at the failure of unit 2 or the Nth failure of unit 1, whichever occurs first. The expected cost rates of two models are obtained, and optimum replacement numbers N^* that minimize them are discussed analytically. Finally, we introduce an extended model of Model 2 where the system is also replaced at time T. The replacement policy for a system with induced failure was extended to multiunit systems [179, 180]. Furthermore, this policy was extended and applied to age and block replacement policies [181–183] and an inspection policy [184].

The above two models characterize some real systems [54]: The following example is the illustrative from the chemical industry. The system consists of a metal container (unit 2) in which chemical reactions take place and the temperature of the container is controlled by cold water pumped through a pneumatic pump (unit 1). Consider the case where the pump fails, and as a result, the pressure inside can build up and lead to an explosion if the quantity of reacting fluid is high. This situation is modeled by Model 1 with $\alpha_j = \alpha$ for all j and α is the probability that the volume of fluid in the container is high. A different scenario is as follows: Whenever the pump fails, the temperature of the tank rises and the container surface is corroded. As a consequence, the thickness of the container decreases. The damage is the reduction in the wall thickness and it is additive. The container fails when the total reduction in the wall thickness has exceeded some specified limit. This situation is modeled by Model 2. Note that without unit 1 failure, there is no damage to unit 2, and hence, it does not fail. If the container is preventively maintained at time T before failure and is like new, the system corresponds to an extended model. The example of a brake pad and disc rotor of an automobile was given [185].

4.2.1 Model 1: Induced Failure

Whenever unit 1 fails, it acts as a shock to induce an instantaneous failure of unit 2 with a certain probability. Let α_j denote the probability that unit 2 fails at the jth failure of unit 1. It is assumed that $0 \equiv \alpha_0 < \alpha_1 \le \alpha_2 \le \cdots \le \alpha_j \le \cdots < 1$. The system is replaced at the failure of unit 2 or at the Nth $(N = 1, 2, \cdots)$ failure of unit 1, whichever occurs first. The system is assumed to be replaced at unit 2 failure, when it fails at the Nth failure of unit 1. The probability that the system is replaced at the Nth failure of unit 1 is

$$(1 - \alpha_1)(1 - \alpha_2) \cdots (1 - \alpha_N), \tag{4.28}$$

and the probability that it is replaced at the failure of unit 2 is

$$\sum_{j=1}^{N} (1 - \alpha_1)(1 - \alpha_2) \cdots (1 - \alpha_{j-1})\alpha_j. \tag{4.29}$$

Note that $(4.28) + (4.29) = 1$.

Because the probability that j failures of unit 1 occur exactly in $[0, t]$ is given by $p_j(t) \equiv \{[H(t)]^j / j!\} e^{-H(t)}$ $(j = 0, 1, 2, \cdots)$ and

$$\int_0^\infty t\, p_{j-1}(t)h(t)\, dt = \int_0^\infty t e^{-H(t)}\, d\left\{ \frac{[H(t)]^j}{j!} \right\}$$

$$= \int_0^\infty t\, p_j(t)h(t)\, dt - \int_0^\infty p_j(t)\, dt,$$

the mean time to replacement is

$$(1 - \alpha_1) \cdots (1 - \alpha_N) \int_0^\infty t\, p_{N-1}(t)h(t)\, dt$$

$$+ \sum_{j=1}^{N} (1 - \alpha_1) \cdots (1 - \alpha_{j-1})\alpha_j \int_0^\infty t\, p_{j-1}(t)h(t)\, dt$$

$$= \sum_{j=0}^{N-1} (1 - \alpha_0)(1 - \alpha_1) \cdots (1 - \alpha_j) \int_0^\infty p_j(t)\, dt. \tag{4.30}$$

The expected number of unit 1 failures before replacement is

$$(N - 1)(1 - \alpha_1) \cdots (1 - \alpha_N) + \sum_{j=1}^{N} (j - 1)(1 - \alpha_1) \cdots (1 - \alpha_{j-1})\alpha_j$$

$$= \sum_{j=1}^{N-1} (1 - \alpha_1) \cdots (1 - \alpha_j), \tag{4.31}$$

where $\sum_1^0 \equiv 0$. Note that we do not include the number of the jth failure in (4.31) when the system is replaced at the jth failure of unit 1.

Let c_1 be the cost of unit 1 failure, c_2 be the replacement cost at the Nth failure of unit 1, and c_3 be the replacement cost at the failure of unit 2 with $c_3 > c_2 > c_1$. Then, the expected cost rate is, from (4.28)–(4.31),

$$C_1(N) = \frac{c_1 \sum_{j=1}^{N-1} A_j + c_3 - (c_3 - c_2) A_N}{\sum_{j=0}^{N-1} A_j \int_0^\infty p_j(t)\, dt} \quad (N = 1, 2, \cdots), \qquad (4.32)$$

where $A_j \equiv (1 - \alpha_0)(1 - \alpha_1) \cdots (1 - \alpha_j)$ $(j = 0, 1, 2, \cdots)$.

We seek an optimum number N^* that minimizes $C_1(N)$ in (4.32). From the inequality $C_1(N + 1) \geq C_1(N)$,

$$c_1 \left[\frac{\sum_{j=0}^{N-1} A_j \int_0^\infty p_j(t)\, dt}{\int_0^\infty p_N(t)\, dt} - \sum_{j=1}^{N-1} A_j \right]$$

$$+ (c_3 - c_2) \left[\frac{A_N - A_{N+1}}{A_N \int_0^\infty p_N(t)\, dt} \sum_{j=0}^{N-1} A_j \int_0^\infty p_j(t)\, dt + A_N \right] \geq c_3$$

$$(N = 1, 2, \ldots). \qquad (4.33)$$

Denoting the left-hand side of (4.33) by $Q_1(N)$,

$$Q_1(N+1) - Q_1(N) = \sum_{j=0}^{N} A_j \int_0^\infty p_j(t)\, dt \left\{ c_1 \left[\frac{1}{\int_0^\infty p_{N+1}(t)\, dt} - \frac{1}{\int_0^\infty p_N(t)\, dt} \right] \right.$$

$$\left. + (c_3 - c_2) \left[\frac{A_{N+1} - A_{N+2}}{A_{N+1} \int_0^\infty p_{N+1}(t)\, dt} - \frac{A_N - A_{N+1}}{A_N \int_0^\infty p_N(t)\, dt} \right] \right\}.$$

Suppose that either of α_j or $h(t)$ is strictly increasing. Then, from (1.29), if $h(t)$ is strictly increasing, then $\int_0^\infty p_j(t) dt$ is strictly decreasing in j to $1/h(\infty)$, where $h(\infty) \equiv \lim_{t \to \infty} h(t)$, and if α_j is strictly increasing, then $(A_N - A_{N+1})/A_N$ is also strictly increasing. Thus, $Q_1(N)$ is strictly increasing in N, and hence, an optimum number N^* is given by a unique minimum such that $Q_1(N) \geq c_3$.

Example 4.3. Suppose that α_j is constant, *i.e.*, $\alpha_j \equiv \alpha$ $(0 < \alpha < 1)$ and $A_j \equiv (1 - \alpha)^j$ $(j = 0, 1, 2, \cdots)$. Then, (4.33) is rewritten as

$$\frac{\sum_{j=0}^{N-1} \alpha(1-\alpha)^j \int_0^\infty p_j(t)\, dt}{\int_0^\infty p_N(t)\, dt} + (1-\alpha)^N \geq \frac{c_1 + \alpha(c_3 - c_1)}{c_1 + \alpha(c_3 - c_2)} \quad (N = 1, 2, \cdots).$$

$$(4.34)$$

If $h(t)$ is strictly increasing, then the left-hand side $Q_1(N)$ of (4.34) is also strictly increasing, and

$$\lim_{N \to \infty} Q_1(N) = \alpha h(\infty) \int_0^\infty e^{-\alpha H(t)}\, dt.$$

Table 4.3. Optimum number N^* to minimize $C_1(N)$ when $\alpha = 0.1$

$(c_3 - c_2)/c_1$	c_2/c_1					
	2	3	5	10	20	50
1	1	2	4	10	24	95
2	1	2	4	9	21	83
5	1	2	3	7	16	58
10	1	1	3	5	12	38
20	1	1	2	4	7	22
50	1	1	1	2	4	10

Thus, if

$$\alpha h(\infty) \int_0^\infty e^{-\alpha H(t)}\, dt > \frac{c_1 + \alpha(c_3 - c_1)}{c_1 + \alpha(c_3 - c_2)}, \tag{4.35}$$

then a finite N^* is given by a unique minimum number that satisfies (4.34).

When $h(t) = 2t$, i.e., $p_j(t) = [(t^2)^j/j!]e^{-t^2}$, $h(t)$ is strictly increasing to ∞. Thus, there exists a unique minimum N^* that satisfies (4.34). Table 4.3 presents the optimum number N^* for $(c_3 - c_2)/c_1 = 1, 2, 5, 10, 20, 50$ and $c_2/c_1 = 2, 3, 5, 10, 20, 50$ when $\alpha = 0.1$. In this case, because $\int_0^\infty p_0(t)dt = \sqrt{\pi}/2$ and $\int_0^\infty p_1(t)dt = \sqrt{\pi}/4$, if $0.1[(c_3 - c_2)/c_1] \geq (c_2/c_1) - 2$, then $N^* = 1$. ∎

Example 4.4. Suppose that $h(t) = \lambda$, i.e., unit 1 failures occur in a Poisson process with rate λ. Then, (4.33) is

$$\alpha_{N+1} \sum_{j=0}^{N-1} A_j + A_N \geq \frac{c_3 - c_1}{c_3 - c_2} \quad (N = 1, 2, \cdots). \tag{4.36}$$

If α_j is strictly increasing in j, where $\alpha_\infty \equiv \lim_{j \to \infty} \alpha_j$ that might be 1, then the left-hand side of (4.36) is also strictly increasing, and

$$Q_1(\infty) \equiv \lim_{N \to \infty} Q_1(N) = \alpha_\infty \sum_{j=0}^\infty A_j.$$

Thus, if $Q_1(\infty) > (c_3 - c_1)/(c_3 - c_2)$, then a finite N^* is a unique minimum that satisfies (4.36). In addition, it is easily proved that

$$\alpha_{N+1} \sum_{j=0}^{N-1} A_j + A_N > \alpha_{N+1} + A_1 \quad (N = 2, 3, \cdots),$$

because

Table 4.4. Optimum number N^* to minimize $C_1(N)$ when $\alpha_j = 1 - (0.9)^j$

$(c_3 - c_2)/c_1$	c_2/c_1					
	2	3	5	10	20	50
1	6	13	∞	∞	∞	∞
2	4	6	13	∞	∞	∞
5	2	3	5	11	∞	∞
10	2	2	3	6	12	∞
20	1	2	2	4	6	20
50	1	1	1	2	3	6

$$\alpha_{N+1} \sum_{j=1}^{N-1} A_j - (A_1 - A_N) = \sum_{j=1}^{N-1} (\alpha_{N+1} A_j + A_{j+1} - A_j)$$

$$= \sum_{j=1}^{N-1} A_j (\alpha_{N+1} - \alpha_{j+1}) > 0.$$

Therefore, if $\alpha_\infty + 1 - \alpha_1 \geq (c_3 - c_1)/(c_3 - c_2)$, then a finite N^* exists.

When $\alpha_j \equiv 1 - \alpha^j$, if a finite N^* exists, then it is given by a unique minimum that satisfies

$$(1 - \alpha^{N+1}) \sum_{j=0}^{N-1} \alpha^{j(j+1)/2} + \alpha^{N(N+1)/2} \geq \frac{c_3 - c_1}{c_3 - c_2} \quad (N = 1, 2, \cdots).$$

Table 4.4 presents the optimum number N^* for $(c_3 - c_2)/c_1 = 1, 2, 5, 10,$ 20, 50 and $c_2/c_1 = 2, 3, 5, 10, 20, 50$ when $\alpha = 0.9$. The optimum N^* increases with c_2/c_1 and decreases with $(c_3 - c_2)/c_1$. Because $\sum_{j=0}^{\infty}(0.9)^{j(j+1)/2} < 3.92$, if $(c_3 - c_1)/(c_3 - c_2) \geq 3.92$, i.e., $c_2/c_1 \geq 1 + 2.92[(c_3 - c_2)/c_1]$, then $N^* = \infty$. If $0.09[(c_3 - c_2)/c_1] \geq (c_2/c_1) - 1$, then $N^* = 1$. ∎

4.2.2 Model 2: Shock Damage

Whenever unit 1 fails, it acts as some shock to unit 2 and causes an amount of damage with distribution $G(x)$ to unit 2. The total damage is additive and unit 2 fails whenever it has exceeded a failure level K. The system is replaced at the failure of unit 2 or at the Nth failure of unit 1, whichever occurs first.

The probability that the system is replaced at the Nth failure of unit 1 is $G^{(N)}(K)$, where $G^{(j)}(x)$ $(j = 1, 2, \cdots)$ is the j-fold Stieltjes convolution of $G(x)$ with itself and $G^{(0)}(x) \equiv 1$ for $x \geq 0$. Thus, the mean time to replacement is, from (3.5),

$$\sum_{j=0}^{N-1} G^{(j)}(K) \int_0^{\infty} p_j(t)\, dt, \tag{4.37}$$

and the expected number of unit 1 failures before replacement is

$$(N-1)G^{(N)}(K) + \sum_{j=1}^{N-1}(j-1)[G^{(j-1)}(K) - G^{(j)}(K)] = \sum_{j=1}^{N-1} G^{(j)}(K). \quad (4.38)$$

Therefore, the expected cost rate is, from (4.37) and (4.38),

$$C_2(N) = \frac{c_1 \sum_{j=1}^{N-1} G^{(j)}(K) + c_3 - (c_3 - c_2)G^{(N)}(K)}{\sum_{j=0}^{N-1} G^{(j)}(K) \int_0^\infty p_j(t)\,dt} \quad (N = 1, 2, \cdots), \quad (4.39)$$

where c_k $(k = 1, 2, 3)$ are the same costs as those for Model 1. In particular, when K goes to infinity,

$$C_2(N) = \frac{c_1(N-1) + c_2}{\sum_{j=0}^{N-1} \int_0^\infty p_j(t)\,dt}, \quad (4.40)$$

that agrees with (4.25) of [1], and it is the expected cost rate of the replacement at the Nth failure.

We seek an optimum number N^* that minimizes $C_2(N)$ in (4.39). From the inequality $C_2(N+1) \geq C_2(N)$,

$$c_1 \left[\frac{1}{\int_0^\infty p_N(t)\,dt} \sum_{j=0}^{N-1} G^{(j)}(K) \int_0^\infty p_j(t)\,dt - \sum_{j=1}^{N-1} G^{(j)}(K) \right]$$

$$+ (c_3 - c_2) \left[\frac{G^{(N)}(K) - G^{(N+1)}(K)}{G^{(N)}(K) \int_0^\infty p_N(t)\,dt} \sum_{j=0}^{N-1} G^{(j)}(K) \int_0^\infty p_j(t)\,dt + G^{(N)}(K) \right]$$

$$\geq c_3 \qquad\qquad (N = 1, 2, \ldots). \quad (4.41)$$

Denoting the left-hand side of (4.41) by $Q_2(N)$,

$$Q_2(N+1) - Q_2(N)$$

$$= \sum_{j=0}^{N} G^{(j)}(K) \int_0^\infty p_j(t)\,dt \left\{ c_1 \left[\frac{1}{\int_0^\infty p_{N+1}(t)\,dt} - \frac{1}{\int_0^\infty p_N(t)\,dt} \right] \right.$$

$$\left. + (c_3 - c_2) \left[\frac{G^{(N+1)}(K) - G^{(N+2)}(K)}{G^{(N+1)}(K) \int_0^\infty p_{N+1}(t)\,dt} - \frac{G^{(N)}(K) - G^{(N+1)}(K)}{G^{(N)}(K) \int_0^\infty p_N(t)\,dt} \right] \right\}.$$

Suppose that either of $[G^{(N)}(K) - G^{(N+1)}(K)]/G^{(N)}(K)$ or $h(t)$ is strictly increasing. Then, $Q_2(N)$ is also strictly increasing in N, and hence, an optimum number N^* is given by a unique minimum that satisfies (4.41).

Example 4.5. Suppose that $G(x) = 1 - e^{-\mu x}$ and $G^{(j)}(K) = \sum_{i=j}^\infty [(\mu K)^i / i!] e^{-\mu K}$. Then, from Example 2.2 of Chapter 2,

$$\frac{G^{(N+1)}(K)}{G^{(N)}(K)} = \frac{\sum_{j=N+1}^{\infty}[(\mu K)^j/j!]}{\sum_{j=N}^{\infty}[(\mu K)^j/j!]}$$

is decreasing in N from $1 - e^{-\mu K}$ to 0. Furthermore,

$$\lim_{N\to\infty} Q_2(N) = (c_3 - c_2 + c_1)h(\infty)\sum_{j=0}^{\infty} G^{(j)}(K)\int_0^{\infty} p_j(t)\,\mathrm{d}t - c_1\mu K.$$

Thus, if

$$h(\infty)\sum_{j=0}^{\infty} G^{(j)}(K)\int_0^{\infty} p_j(t)\,\mathrm{d}t > \frac{c_3 + c_1\mu K}{c_3 - c_2 + c_1},$$

then a finite N^* is given by a unique minimum number that satisfies (4.41). In addition, when $h(t) = \lambda$, $h(\infty)\int_0^{\infty} p_j(t)\mathrm{d}t = 1$ and $\sum_{j=0}^{\infty} G^{(j)}(K) = 1 + \mu K$, and hence, if $\mu K > (c_2 - c_1)/(c_3 - c_2)$, then a finite N^* exists uniquely. ∎

4.2.3 Modified Models

(1) Case of Renewal Process

If unit 1 fails, then it is replaced with a new one, that is, unit 1 failures occur in a renewal process with mean interval $1/\lambda$. Then, the expected cost rate of Model 1 is, from (4.32),

$$\frac{C_1(N)}{\lambda} = \frac{c_1\sum_{j=1}^{N-1} A_j + c_3 - (c_3 - c_2)A_N}{\sum_{j=0}^{N-1} A_j} \qquad (N = 1, 2, \cdots). \qquad (4.42)$$

Thus, the optimum number N^* that minimizes $C_1(N)$ has been derived in Example 4.4

Similarly, the expected cost rate of Model 2 is, from (4.39),

$$\frac{C_2(N)}{\lambda} = \frac{c_1\sum_{j=1}^{N-1} G^{(j)}(K) + c_3 - (c_3 - c_2)G^{(N)}(K)}{\sum_{j=0}^{N-1} G^{(j)}(K)} \qquad (N = 1, 2, \cdots).$$

$$(4.43)$$

Thus, the optimum number N^* that minimizes $C_2(N)$ is derived in **(2)** of Section 3.2, by replacing $c_K = c_3 - c_1$ and $c_N = c_2 - c_1$.

(2) Replacement at Time T and Shock N for Model 2

Consider an extended replacement policy for Model 2 where the system is replaced at time T, at the failure of unit 2, or at the Nth failure of unit 1, whichever occurs first.

The probability that the system is replaced at time T is

$$\sum_{j=0}^{N-1} p_j(T) G^{(j)}(K),\qquad(4.44)$$

the probability that it is replaced at the Nth failure of unit 1 is

$$\sum_{j=N}^{\infty} p_j(T) G^{(N)}(K),\qquad(4.45)$$

and the probability that it is replaced at the failure of unit 2 is

$$\sum_{j=0}^{N-1} p_j(T)[1 - G^{(j)}(K)] + \sum_{j=N}^{\infty} p_j(T)[1 - G^{(N)}(K)]$$

$$= \sum_{j=1}^{N} [G^{(j-1)}(K) - G^{(j)}(K)] \sum_{i=j}^{\infty} p_i(T).\qquad(4.46)$$

It is clearly seen that $(4.44) + (4.45) + (4.46) = 1$. The mean time to replacement is

$$T \sum_{j=0}^{N-1} p_j(T) G^{(j)}(K) + G^{(N)}(K) \int_0^T t\, p_{N-1}(t) h(t)\, dt$$

$$+ \sum_{j=1}^{N} [G^{(j-1)}(K) - G^{(j)}(K)] \int_0^T t\, p_{j-1}(t) h(t)\, dt$$

$$= \sum_{j=0}^{N-1} G^{(j)}(K) \int_0^T p_j(t)\, dt,\qquad(4.47)$$

and the expected number of unit 1 failures before replacement is

$$\sum_{j=0}^{N-1} j p_j(T) G^{(j)}(K) + (N-1) \sum_{j=N}^{\infty} p_j(T) G^{(N)}(K)$$

$$+ \sum_{j=1}^{N} (j-1)[G^{(j-1)}(K) - G^{(j)}(K)] \sum_{i=j}^{\infty} p_i(T)$$

$$= \sum_{j=1}^{N-1} G^{(j)}(K) \sum_{i=j}^{\infty} p_i(T).\qquad(4.48)$$

Therefore, the expected cost rate is, from (4.44)–(4.48),

$$C(T,N) = \frac{\begin{array}{l} c_1 \sum_{j=1}^{N-1} G^{(j)}(K) \sum_{i=j}^{\infty} p_i(T) + c_2 G^{(N)}(K) \sum_{j=N}^{\infty} p_j(T) \\ + c_3 \sum_{j=1}^{N-1} [G^{(j-1)}(K) - G^{(j)}(K)] \sum_{i=j}^{\infty} p_i(T) \\ + c_4 \sum_{j=0}^{N-1} G^{(j)}(K) p_j(T) \end{array}}{\sum_{j=0}^{N-1} G^{(j)}(K) \int_0^T p_j(t)\, dt},\qquad(4.49)$$

where c_1 = cost of one unit failure, c_2 =replacement cost at the Nth failure of unit 1, c_3 = replacement cost at the failure of unit 2, and c_4 = replacement cost at time T. In particular, when T goes to infinity, $C(T, N)$ agrees with $C_2(N)$ in (4.39)

On the other hand, when N goes to infinity and unit 1 failures occur in a Poisson process with rate λ, i.e., $p_j(t) = [(\lambda t)^j / j!]e^{-\lambda t}$ ($j = 0, 1, 2, \cdots$), the expected cost rate is simplified as

$$C(T) \equiv \lim_{N \to \infty} C(T, N)$$

$$= \frac{c_3 - c_1 - (c_3 - c_1 - c_4) \sum_{j=0}^{\infty} G^{(j)}(K)p_j(T)}{(1/\lambda) \sum_{j=0}^{\infty} G^{(j)}(K) \sum_{i=j+1}^{\infty} p_i(T)} + \lambda c_1. \qquad (4.50)$$

Thus, the optimum problem of minimizing $C(T)$ corresponds to that of minimizing $C_1(T)$ in (3.11) when $p_j(t) = F^{(j)}(t) - F^{(j+1)}(t)$.

(3) Replacement at Time T and Damage Z

Consider the replacement model where the system is replaced before failure of unit 2 when its total damage has exceeded a threshold level Z ($0 \le Z \le K$) without replacing at the Nth failure of unit 1 in (2). It is supposed that the system is replaced at time T, at the failure of unit 2, or at damage Z, whichever occurs first [185].

The probability that the system is replaced at time T is

$$\sum_{j=0}^{\infty} G^{(j)}(Z)p_j(T), \qquad (4.51)$$

the probability that it is replaced at damage Z, i.e., when the total damage has exceeded Z and is less than K, is

$$\sum_{j=0}^{\infty} \int_0^Z [G(K - x) - G(Z - x)] \, dG^{(j)}(x) \sum_{i=j+1}^{\infty} p_i(T), \qquad (4.52)$$

and the probability that it is replaced at the failure of unit 2, i.e., when the total damage has exceeded a failure level K, is

$$\sum_{j=0}^{\infty} \int_0^Z [1 - G(K - x)] \, dG^{(j)}(x) \sum_{i=j+1}^{\infty} p_i(T). \qquad (4.53)$$

Note that (4.51) + (4.52) + (4.53) = 1. The mean time to replacement is

$$T \sum_{j=0}^{\infty} G^{(j)}(Z)p_j(T) + \sum_{j=0}^{\infty} \int_0^Z [1 - G(Z - x)] \, dG^{(j)}(x) \int_0^T t\, p_j(t)h(t) \, dt$$

$$= \sum_{j=0}^{\infty} G^{(j)}(Z) \int_0^T p_j(t) \, dt, \qquad (4.54)$$

and the expected number of unit 1 failures before replacement is

$$\sum_{j=0}^{\infty} jG^{(j)}(Z)p_j(T) + \sum_{j=0}^{\infty} j \int_0^Z [1 - G(Z - x)] \, dG^{(j)}(x) \sum_{i=j+1}^{\infty} p_i(T)$$

$$= \sum_{j=1}^{\infty} G^{(j)}(Z) \sum_{i=j}^{\infty} p_i(T). \tag{4.55}$$

Denoting that c_2 is the replacement cost at damage Z and the other costs are the same ones as those of (4.49), the expected cost rate is, from (4.51)–(4.55),

$$C(T, Z) = \frac{\begin{aligned} &c_1 \sum_{j=1}^{\infty} G^{(j)}(Z) \sum_{i=j}^{\infty} p_i(T) \\ &+ c_2 \sum_{j=0}^{\infty} \int_0^Z [G(K - x) - G(Z - x)] \, dG^{(j)}(x) \sum_{i=j+1}^{\infty} p_i(T) \\ &+ c_3 \sum_{j=0}^{\infty} \int_0^Z [1 - G(K - x)] \, dG^{(j)}(x) \sum_{i=j+1}^{\infty} p_i(T) \\ &+ c_4 \sum_{j=0}^{\infty} G^{(j)}(Z)p_j(T) \end{aligned}}{\sum_{j=0}^{\infty} G^{(j)}(Z) \int_0^T p_j(t) \, dt}.$$

$$\tag{4.56}$$

It is clearly seen that $C(T, Z)$, as $Z \to K$, is equal to $C(T, N)$ in (4.49), as $N \to \infty$. There do not exist both T^* $(0 < T^* < \infty)$ and Z^* $(0 < Z^* < K)$ that minimize the expected cost rate $C(T, Z)$ as shown in (2) of Section 3.3.

Suppose that the system is replaced before failure only at damage Z and $p_j(t) = [(\lambda t)^j / j!]e^{-\lambda t}$ $(j = 0, 1, 2, \cdots)$. Then, the expected cost rate is

$$C(Z) \equiv \lim_{T \to \infty} C(T, Z)$$

$$= \frac{(c_3 - c_2 + c_1)M_G(Z) + c_3 - (c_3 - c_2)[G(K) + \int_0^Z G(K - x) \, dM_G(x)]}{[1 + M_G(Z)]/\lambda},$$

$$\tag{4.57}$$

where $M_G(x) \equiv \sum_{j=1}^{\infty} G^{(j)}(x)$. When $c_1 = 0$, this corresponds to the expected cost rate in (3.24).

5

Periodic Replacement Policies

When we consider large and complex systems that consist of many different kinds of units, we should make the planned replacement or preventive maintenance at periodic times, and make some minimal repair at failures between replacements. This policy is called *periodic replacement with minimal repair at failures* [66], where minimal repair means that the failure rate remains undisturbed by any repair of failures. A unit is inspected and replaced periodically at planned times nT $(n = 1, 2, \cdots)$. This replacement policy is commonly used with complex systems such as computers, airplanes, and large production systems. Their theoretical results were extensively summarized [1].

This chapter applies the periodic replacement to a cumulative damage model where shocks occur in a renewal process and the total damage due to shocks is additive. This periodic replacement was considered, and optimum policies that minimize the expected costs under suitable conditions were discussed [186–189].

We have already derived the failure distribution $\Phi(t)$ in (2.9) of a unit with cumulative damage. Substituting $\Phi(t)$ in standard replacements such as age replacement, block replacement, and periodic inspection, it is shown in Section 5.1 that these replacement policies can be applied to a cumulative damage model. In Section 5.2, the amount of total damage is checked only at periodic times nT, and a unit is replaced before failure at a planned time NT. The expected cost rate is obtained and an optimum N^* that minimizes it is derived [190]. It has been assumed in all models until now that a unit is always replaced at failures. Section 5.3 considers the cumulative damage model where a unit suffers some damage caused by both shock and inspection [191]. In Section 5.4, we apply the periodic replacement with minimal repair at failures to a cumulative damage model [55]. It is assumed that a unit fails with probability $p(x)$ when that total damage becomes x at shocks and the total damage is not unchanged by any minimal repair at failures. The expected cost rate is obtained, and an optimum planned time T^*, shock number N^* and damage level Z^* that minimize it are discussed analytically. Furthermore, in Section 5.5, we consider modified models where a unit is replaced at the next shock, when the total operating time has exceeded a planned time T and the total damage has exceeded a damage level Z. Numerical examples to understand these models and methods easily are given in some sections.

5.1 Basic Replacement Models

Suppose that the failure distribution $\Phi(t)$ of a unit with a failure level K is given in (2.9), where $\overline{\Phi} \equiv 1 - \Phi$. Then, using the theory of replacement policies [1], we have the following expected cost rates: A unit is replaced with a new one at a planned time T ($0 < T \le \infty$) or at failure, $i.e.$, when the total damage has exceeded a failure level K, whichever occurs first. This is called an *age replacement policy* and its expected cost rate is, from (3.4) of [1],

$$C_1(T) = \frac{(c_K - c_T)\Phi(T) + c_T}{\int_0^T \overline{\Phi}(t)\,\mathrm{d}t},\tag{5.1}$$

where cost c_K is incurred for the replacement of a failed unit and cost c_T ($< c_K$) is incurred for the replacement of a nonfailed unit at time T.

A unit is replaced with a new one at periodic times nT ($n = 1, 2, \cdots$) and is also replaced at each failure between periodic replacements. This is called a *block replacement* and its expected cost rate is, from (5.1) of [1],

$$C_2(T) = \frac{1}{T}\left[c_K M_\Phi(T) + c_T\right],\tag{5.2}$$

where c_K is the cost of replacement at each failure, c_T is the cost of the planned replacement, $M_\Phi(t) \equiv \sum_{n=1}^{\infty} \Phi^{(n)}(t)$ is a renewal function of a failure distribution $\Phi(t)$, and $\Phi^{(n)}(t)$ is the n-fold Stieltjes convolution of $\Phi(t)$ and $\Phi^{(0)}(t) \equiv 1$ for $t \ge 0$.

Furthermore, when a unit fails between periodic replacements, it remains in a failed state and is replaced only at a planned time T. Then, the expected cost rate is, from (5.10) of [1],

$$C_3(T) = \frac{1}{T}\left[c_D \int_0^T \Phi(t)\,\mathrm{d}t + c_T\right],\tag{5.3}$$

where c_D is the downtime cost per unit of time for the time elapsed between a failure and its replacement. Optimum policies that minimize $C_k(T)$ ($k = 1, 2, 3$) were discussed analytically for a general failure distribution [1].

Finally, any failure is detected only through inspection. A unit is checked at periodic times nT ($n = 1, 2, \cdots$), its failure is always detected at the next checking time, and it is replaced. This is called an *inspection policy* with replacement, and the total expected cost until replacement is, from (8.1) of [1],

$$C_4(T) = \sum_{n=0}^{\infty} \int_{nT}^{(n+1)T} \left\{c_T(n+1) + c_D\left[(n+1)T - t\right]\right\}\,\mathrm{d}\Phi(t) + c_K,\tag{5.4}$$

where c_T is the cost of one check at time nT, c_D is the loss cost per unit of time for the time elapsed between a failure and its detection, and c_K is the replacement cost of a failed unit.

Example 5.1. Suppose that shocks occur in a Poisson process, each damage due to shocks and a failure level K are exponential, *i.e.*, $F(t) = 1 - \mathrm{e}^{-\lambda t}$, $G(x) = 1 - \mathrm{e}^{-\mu x}$, and $L(x) = 1 - \mathrm{e}^{-\theta x}$. Then, from Example 2.3,

$$\Phi(t) = 1 - \exp\left(-\frac{\lambda \theta t}{\mu + \theta}\right).$$

The total expected cost of an inspection policy is, from (5.4),

$$C_4(T) = \frac{c_T + c_D T}{1 - \mathrm{e}^{-\lambda \theta T/(\mu+\theta)}} - \frac{c_D\,(\mu + \theta)}{\lambda \theta} + c_K.$$

Thus, an optimum checking time T^* to minimize $C_4(T)$ is given by a unique solution that satisfies

$$\mathrm{e}^{\lambda \theta T/(\mu+\theta)} - \left(1 + \frac{\lambda \theta T}{\mu + \theta}\right) = \frac{c_T \lambda \theta}{c_D(\mu + \theta)}$$

and it is approximately

$$\widetilde{T} = \sqrt{\frac{2c_T(\mu + \theta)}{c_D \lambda \theta}},$$

and $T^* < \widetilde{T}$.

Next, suppose that $\Phi(t)$ is an exponential distribution with mean K/a $(a > 0)$ from Section 2.4, *i.e.*, when Y is the time to failure, $aE\{Y\} = K$ and $\Phi(t) = 1 - \mathrm{e}^{-at/K}$. Then, the total expected cost is

$$C_4(T) = \frac{c_T + c_D T}{1 - \mathrm{e}^{-aT/K}} - \frac{c_D K}{a} + c_K.$$

An optimum T^* satisfies

$$\mathrm{e}^{aT/K} - \left(1 + \frac{aT}{K}\right) = \frac{c_T a}{c_D K},$$

and it is approximately

$$\widetilde{T} = \sqrt{\frac{2c_T K}{c_D a}},$$

and $T^* < \widetilde{T}$. It is clearly seen that T^* decreases, as parameter a increases. This represents the continuous wear model in which the failure time is distributed exponentially and its mean time is $E\{Y\} = K/a$.

When shocks occur in a Poisson distribution with mean $1/\lambda$ and a unit fails at shock n, $\Phi(t)$ has a gamma distribution in (1.23), i.e., $\Phi(t) = \sum_{i=n}^{\infty}[(\lambda t)^i/i!]e^{-\lambda t}$ $(n = 1, 2, \dots)$. In this case, the total expected cost is

$$C_4(T) = (c_T + c_D T)\sum_{j=0}^{\infty}\sum_{i=0}^{n-1}\frac{(\lambda j T)^i}{i!}e^{-\lambda j T} - \frac{n c_D}{\lambda} + c_K. \quad \blacksquare$$

Similar replacement policies when $\Phi(t)$ is a gamma distribution were considered [192, 193]. This is called a continuous wear process under discrete monitoring by inspection, that is one of conditioned maintenance policies as shown in Section 6.1. Multicritical levels of preventive maintenances for a failure level K were proposed, and the optimum policies for several systems were discussed [194–196].

5.2 Discrete Replacement Models

Each amount W_n $(n = 1, 2, \cdots)$ of damage to a unit is measured only at planned times nT $(n = 1, 2, \cdots)$ for a given T $(0 < T < \infty)$ and has an identical distribution $G(x) \equiv \Pr\{W_n \le x\}$ between periodic times. The unit fails only at time nT, and is replaced at time NT or at failure, whichever occurs first. Because the mean time to replacement is

$$\sum_{n=0}^{N-1}[(n+1)T][G^{(n)}(K) - G^{(n+1)}(K)] + (NT)G^{(N)}(K) = T\sum_{n=0}^{N-1}G^{(n)}(K),$$

the expected cost rate is

$$C_1(N) = \frac{c_K - (c_K - c_N)G^{(N)}(K)}{T\sum_{n=0}^{N-1}G^{(n)}(K)} \quad (N = 1, 2, \cdots), \tag{5.5}$$

where c_K is the replacement cost at failure and c_N $(< c_K)$ is the replacement cost at time NT. Thus, this corresponds to the same replacement model with a shock number N in (2) of Section 3.2, by replacing $1/\lambda$ with T. The replacement policy where the unit is replaced before failure at damage Z has been already taken up in (3) of Section 3.2.

Next, suppose that shocks occur continuously and the total damage is proportional to an operating time, i.e., $Z(t) = at$ $(a > 0)$. In this case, if a failure level K is a random variable with a continuous distribution $L(x)$ defined in (2) of Section 2.5, the probability that the unit fails at time nT is $\Pr\{naT \ge K\} = L(naT)$. Thus, the probability that the unit fails until time NT is

$$\sum_{n=1}^{N}L(naT)\prod_{i=0}^{n-1}\overline{L}(iaT), \tag{5.6}$$

and the probability that it does not fail until time NT is

$$\prod_{n=1}^{N} \bar{L}(naT),$$ (5.7)

where $\bar{L}(x) \equiv 1 - L(x)$. Note that $(5.6) + (5.7) = 1$. The mean time to replacement is

$$\sum_{n=1}^{N}(nT)L(naT)\prod_{i=0}^{n-1}\bar{L}(iaT) + (NT)\prod_{n=1}^{N}\bar{L}(naT) = T\sum_{n=0}^{N-1}\left[\prod_{i=0}^{n}\bar{L}(iaT)\right],$$ (5.8)

and hence, the mean time $E\{Y\}$ to failure is

$$E\{Y\} = T\sum_{n=0}^{\infty}\left[\prod_{i=0}^{n}\bar{L}(iaT)\right].$$

Therefore, the expected cost rate is, from (5.7) and (5.8),

$$C_2(N) = \frac{c_K - (c_K - c_N)\prod_{n=1}^{N}\bar{L}(naT)}{T\sum_{n=0}^{N-1}\left[\prod_{i=0}^{n}\bar{L}(iaT)\right]} \quad (N = 1, 2, \cdots).$$ (5.9)

We seek an optimum number N^* that minimizes $C_2(N)$. From the inequality $C_2(N+1) - C_2(N) \geq 0$,

$$L\left((N+1)aT\right)\sum_{n=0}^{N-1}\left[\prod_{i=0}^{n}\bar{L}(iaT)\right] + \prod_{n=1}^{N}\bar{L}(naT) \geq \frac{c_K}{c_K - c_N} \quad (N = 1, 2, \cdots).$$ (5.10)

Letting $Q(N)$ be the left-hand side of (5.10),

$$Q(\infty) \equiv \lim_{N\to\infty} Q(N) = \sum_{n=0}^{\infty}\left[\prod_{i=0}^{n}\bar{L}(iaT)\right] = \frac{E\{Y\}}{T},$$

$$Q(N+1) - Q(N) = [L((N+2)aT) - L((N+1)aT)]\sum_{n=0}^{N}\left[\prod_{i=0}^{n}\bar{L}(iaT)\right] > 0.$$

Thus, $Q(N)$ is strictly increasing to $E\{Y\}/T$ that represents the expected number of periodic times to failure, and hence, we have the optimum replacement policy:

(i) If $E\{Y\}/T > c_K/(c_K - c_N)$, then there exists a finite and unique minimum N^* $(1 \leq N^* < \infty)$ that satisfies (5.10), and its resulting cost rate is

$$\frac{L(N^*aT)}{T(c_K - c_N)} < C_2(N^*) \leq \frac{L((N^*+1)aT)}{T(c_K - c_N)}.$$ (5.11)

(ii) If $E\{Y\}/T \le c_K/(c_K - c_N)$, then $N^* = \infty$, i.e., the unit should be replaced only at failure, and

$$C_2(\infty) \equiv \lim_{N \to \infty} C_2(N)$$

$$= \frac{c_K}{T \sum_{n=0}^{\infty} \left[\prod_{i=0}^{n} \overline{L}(iaT)\right]} = \frac{c_K}{E\{Y\}}. \tag{5.12}$$

In particular, when $L(x) = 1 - e^{-\theta x}$, (5.10) is

$$\left[1 - e^{-a\theta T(N+1)}\right] \sum_{n=0}^{N-1} e^{-a\theta T[n(n+1)/2]} + e^{-a\theta T[N(N+1)/2]} \ge \frac{c_K}{c_K - c_N}, \tag{5.13}$$

and

$$Q(\infty) = \sum_{n=0}^{\infty} e^{-a\theta T[n(n+1)/2]}. \tag{5.14}$$

Example 5.2. Suppose that a failure level K is normally distributed with mean k and standard deviation σ, and furthermore, $aT = 1$, i.e., $\overline{L}(naT) = [1/(\sqrt{2\pi}\sigma)] \int_n^{\infty} \exp\left[-(x-k)^2/(2\sigma^2)\right] dx$ $(n = 0, 1, 2, \cdots)$. Then, Table 5.1 presents the optimum replacement number N^* and the mean time $E\{Y\}$ to failure for $k = 10, 20, 50$ and $\sigma = 1, 2, 5, 10$ when $c_K/c_N = 5$.

Another single method of such replacements is to balance the cost of replacement at failure against that at nonfailure, i.e., $c_K \times (5.6) \ge c_N \times (5.7)$. In this case,

$$\prod_{n=1}^{N} \overline{L}(naT) \le \frac{c_K}{c_K + c_N},$$

and a minimum $\widetilde{N}$ to satisfy it is also presented in Table 5.1. This indicates that the values of N^*, $\widetilde{N}$, and $E\{Y\}$ decrease with σ because the variance of a failure level becomes larger. Furthermore, when $\sigma = 1$, the unit should be replaced before failure at 68.2%, 83.9%, 93.5% of the mean failure time for $k = 10, 20, 50$, respectively, and $N^* = k - 3\sigma$ for all k. When σ is small, the approximate $\widetilde{N}$ gives a good upper bound of N^*. It is of interest that $k > E\{Y\}/T > \widetilde{N} > N^*$ for $\sigma \ge 2$. ∎

5.3 Deteriorated Inspection Model

We introduce the replacement policy for the cumulative damage model where a unit is checked at periodic times nT $(n = 1, 2, \ldots)$ [197]. It has been generally

Table 5.1. Comparison of optimum number N^*, approximate value $\tilde{N}$, and mean time $E\{Y\}/T$ to failure when $aT = 1$ and $c_K/c_N = 5$

σ	$k = 10$			$k = 20$			$k = 50$		
	N^*	$\tilde{N}$	$E\{Y\}/T$	N^*	$\tilde{N}$	$E\{Y\}/T$	N^*	$\tilde{N}$	$E\{Y\}/T$
1	7	9	10.27	17	19	20.27	47	49	50.27
2	5	8	9.51	15	18	19.51	44	48	49.51
5	2	4	6.40	8	13	15.99	36	43	45.99
10	1	1	3.91	3	5	9.55	23	33	38.38

assumed that any inspection does not degrade a unit [1]. On the other hand, the inspection policy for a storage system that is degraded with time and at each inspection was proposed [198]. This could be applied to the periodic test of electric equipment in storage [199].

This section considers the cumulative damage model where a unit suffers some damage and deterioration caused by both shocks and inspections and fails when the total damage has exceeded a failure level K (Figure 5.1). A unit is checked to detect a failure at periodic times nT ($n = 1, 2, \ldots$), where T is previously given, *i.e.*, the failure is detected only through inspection. In addition, to prevent failures, a unit is replaced before failure with a new one at a planned time NT.

5.3.1 Expected Cost

Suppose that the number of shocks in $[0, t]$ is $N(t)$, and the probability that j shocks occur in $[0, t]$ is $F_j(t) \equiv \Pr\{N(t) \geq j\}$ defined in Section 3.1. An amount W_j of damage due to the jth shock has an identical distribution $G(x) \equiv \Pr\{W_j \leq x\}$, $G^{(j)}(x)$ is the j-fold Stieltjes convolution of $G(x)$ with itself, and $G^{(0)}(x) \equiv 1$ for $x \geq 0$. Furthermore, the unit is checked at periodic times nT ($n = 1, 2, \ldots$), where the inspection time is negligible, and each inspection causes a constant and nonnegative amount w of damage to the unit. Let $\overline{N}$ denote the upper number of inspections until the unit fails, *i.e.*, $\overline{N} \equiv [K/w]$, where $[x]$ denotes the greatest integer contained in x and $N = \infty$ whenever $w = 0$.

From the assumption that the unit fails when the total damage has exceeded K, the reliability function $\overline{\Phi}(t)$ that it does not fail in time t for $nT < t \leq (n+1)T$ ($n = 0, 1, 2, \ldots, \overline{N}$) is given by

$$\overline{\Phi}(t) \equiv \Pr\left\{ \sum_{j=0}^{N(t)} W_j + nw \leq K \right\} = \sum_{j=0}^{\infty} G^{(j)}(K - nw)[F_j(t) - F_{j+1}(t)]. \quad (5.15)$$

A unit is always replaced at the first inspection when the total damage has exceeded K. To prevent a failure, the unit is also replaced before failure

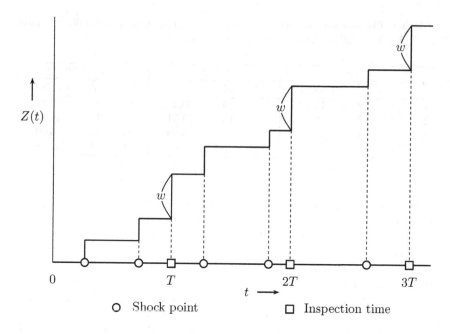

Fig. 5.1. Process for periodic inspection with deteriorated factor w

at the Nth inspection ($N = 1, 2, \ldots, \overline{N}$). Let us introduce three costs given in (5.4). Costs c_K and c_T are incurred for each replacement and inspection, respectively, and c_D is incurred for the time elapsed between a failure and its detection per unit of time. Then, the expected cost until replacement is, from 8.1 of [1] and (5.4),

$$\sum_{n=0}^{N-1} \int_{nT}^{(n+1)T} \{c_T(n+1) + c_D[(n+1)T - t]\}\, d\Phi(t) + c_T N \overline{\Phi}(NT) + c_K$$

$$= (c_T + c_D T) \sum_{n=0}^{N-1} \overline{\Phi}(nT) - c_D \sum_{n=0}^{N-1} \int_{nT}^{(n+1)T} \overline{\Phi}(t)\, dt + c_K, \qquad (5.16)$$

and the mean time to replacement is

$$\sum_{n=0}^{N-1} [(n+1)T] \int_{nT}^{(n+1)T} d\Phi(t) + (NT)\overline{\Phi}(NT) = T \sum_{n=0}^{N-1} \overline{\Phi}(nT), \qquad (5.17)$$

where $\Phi(t) \equiv 1 - \overline{\Phi}(t)$.

Therefore, the expected cost rate is, from (5.16) and (5.17),

$$C(N) = \frac{(c_T + c_D T)\sum_{n=0}^{N-1} \overline{\Phi}(nT) - c_D \sum_{n=0}^{N-1} \int_{nT}^{(n+1)T} \overline{\Phi}(t)\, dt + c_K}{T \sum_{n=0}^{N-1} \overline{\Phi}(nT)}$$

$$(N = 1, 2, \ldots, \overline{N}). \qquad (5.18)$$

5.3.2 Optimum Policy

We find an optimum planned number N^* that minimizes the expected cost rate $C(N)$ in (5.18). Forming the inequality $C(N + 1) \geq C(N)$,

$$\sum_{n=0}^{N-1} \int_{nT}^{(n+1)T} \overline{\Phi}(t)\, dt - \frac{\sum_{n=0}^{N-1} \overline{\Phi}(nT)}{\overline{\Phi}(NT)} \int_{NT}^{(N+1)T} \overline{\Phi}(t)\, dt \geq \frac{c_K}{c_D}$$

$$(N = 1, 2, \ldots \overline{N}). \qquad (5.19)$$

Denoting the left-hand side of (5.19) by $Q(N)$,

$$Q(N+1) - Q(N) = \sum_{n=0}^{N} \overline{\Phi}(nT) \left[\frac{\int_{NT}^{(N+1)T} \overline{\Phi}(t)\, dt}{\overline{\Phi}(NT)} - \frac{\int_{(N+1)T}^{(N+2)T} \overline{\Phi}(t)\, dt}{\overline{\Phi}((N+1)T)} \right]. \quad (5.20)$$

First, prove that if the failure rate of $\Phi(t)$ is strictly increasing, then (5.20) is positive, i.e., $Q(N)$ is strictly increasing in N. From the definition of the failure rate, if the failure rate of $\Phi(t)$ is increasing, then $\overline{\Phi}(t + x)/\overline{\Phi}(t)$ is decreasing in t for any $x > 0$ [1, p. 7]. Thus, because

$$\frac{\int_{NT}^{(N+1)T} \overline{\Phi}(t)\, dt}{\overline{\Phi}(NT)} = \frac{\int_0^T \overline{\Phi}(t + NT)\, dt}{\overline{\Phi}(NT)},$$

we can prove that if the failure rate of $\Phi(t)$ is increasing, then $\overline{\Phi}(t + NT)/\overline{\Phi}(NT)$ is decreasing in NT for any $0 < t < T$, i.e., $\int_{NT}^{(N+1)T} \overline{\Phi}(t)\, dt/\overline{\Phi}(NT)$ is decreasing in N, and hence, $Q(N + 1) - Q(N) \geq 0$.

Therefore, we have the following optimum policy when the failure rate of $\Phi(t)$ is strictly increasing:

(i) If $Q(\overline{N}) > c_K/c_D$, then there exists a unique minimum N^* that satisfies (5.19).
(ii) If $Q(\overline{N}) \leq c_K/c_D$, then $N^* = \overline{N} + 1$, i.e., the unit is always replaced after failure.

Example 5.3.   Suppose that shocks occur in a Poisson process with rate λ and the amount of damage due to each shock has an exponential distribution $(1 - e^{-\mu x})$, that is,

$$F_j(t) = \sum_{i=j}^{\infty} \frac{(\lambda t)^i}{i!} e^{-\lambda t}, \qquad G^{(j)}(x) = \sum_{i=j}^{\infty} \frac{(\mu x)^i}{i!} e^{-\mu x} \quad (j = 0, 1, 2, \ldots),$$

and $\Phi(t)$ in (5.15) is, for $nT < t \leq (n+1)T$ $(n = 0, 1, 2, \ldots, \overline{N})$,

$$\overline{\Phi}(t) = \sum_{j=0}^{\infty} \frac{(\lambda t)^j}{j!} e^{-\lambda t} \sum_{i=j}^{\infty} \frac{[\mu(K - nw)]^i}{i!} e^{-\mu(K - nw)}.$$

Table 5.2. Optimum number N^* and expected cost rate $C(N^*)$ for μw and $\lambda c_K/c_D$ when $\lambda T = 5$, $\lambda = 1$, $c_D = 1$, $c_T = 1$, and $\mu K = 100$

μw	$\overline{N}$	N^*			$C(N^*)$		
		$\lambda c_K/c_D$			$\lambda c_K/c_D$		
		1	5	10	1	5	10
0	∞	15	17	19	0.214	0.264	0.323
1	100	13	15	16	0.216	0.274	0.340
2	50	11	13	14	0.218	0.284	0.359
3	33	10	11	12	0.220	0.295	0.379
4	25	9	10	11	0.222	0.304	0.398
5	20	8	9	10	0.225	0.314	0.416
6	16	8	9	9	0.226	0.323	0.435
7	14	7	8	9	0.229	0.332	0.454
8	12	7	8	8	0.230	0.343	0.480
9	11	6	7	8	0.233	0.350	0.490
10	10	6	7	7	0.234	0.360	0.505

The failure rate of $\overline{\Phi}(t)$ is

$$r(t) \equiv \frac{\Phi'(t)}{\overline{\Phi}(t)} = \frac{\lambda \sum_{j=0}^{\infty} [(\lambda t)^j/j!][G^{(j)}(x) - G^{(j+1)}(x)]}{\sum_{j=0}^{\infty} [(\lambda t)^j/j!]G^{(j)}(x)},$$

where $x \equiv K - nw$. Note from Section 2.3 that $r(t)$ is strictly increasing.

Table 5.2 presents the optimum number N^* and $\overline{N} = [100/\mu w]$ for $\mu w = 0$, 1, ..., 10 and $\lambda c_K/c_D = 1, 5, 10$ when $\lambda T = 5$, $c_T = 1$, $\mu K = 100$, and the resulting cost rate $C(N^*)$ when $\lambda = 1$ and $c_D = 1$. For example, when $\lambda T = 5$, $\mu w = 5$, and $\lambda c_K/c_D = 1$, $N^* = 8$, that is , when shocks occur 5 times a week and the unit fails at about $K/(5/\mu + w) = 10$ weeks, on the average, it should be replaced at 8 weeks. The optimum N^* decreases to 1 with μw. The reason would be that the mean time to replacement greatly decreases with μw. Conversely, $C(N^*)$ slowly increases with μw, because the decrease of the total cost would influence less $C(N^*)$ than the time to failure. It is of interest that $N^* + \mu w$ decreases first, is constant for a while, and increases slowly with μw. ∎

5.4 Replacement with Minimal Repair

It has been assumed in all models that a unit is always replaced at failure. We apply the periodic replacement with minimal repair at failure (Chapter 4 of [1]) to a cumulative damage model.

Consider a cumulative damage model as shown in Section 2.1: Shocks occur in a renewal process with a general distribution $F(t)$ having finite mean $1/\lambda$, and an amount of damage due to each shock has an identical distribution $G(x)$. In this case, the distribution of the total damage $Z(t)$ at time t is given in (2.3). In addition, a unit fails with probability $p(x)$, that is increasing in x from 0 to 1, when the total damage becomes x at shocks, and undergoes only minimal repair at failures, where the total damage remains undisturbed by any minimal repair. To prevent failures, a unit is replaced at a planned time T, at a shock number N, or at a damage level Z, whichever occurs first. Strictly speaking, the policy where a unit is replaced at N or Z is not periodic. However, denoting one cycle from the beginning of operation to the replacement at N or Z, the policy forms a renewal process and the time of each cycle is nearly periodic.

5.4.1 Expected Cost

A unit fails with probability $p(x)$ when the total damage becomes x at each shock in the cumulative damage model and undergoes only minimal repair at failures, $i.e.$, its damage remains undisturbed by minimal repair and its time for minimal repair is negligible. It is assumed that a unit is replaced at time T, at shock N, or at damage Z, whichever occurs first. The probability that the unit is replaced at time T is, from (3.1),

$$P_T = \sum_{j=0}^{N-1} \left[F^{(j)}(T) - F^{(j+1)}(T) \right] G^{(j)}(Z), \tag{5.21}$$

the probability that it is replaced at shock N is, from (3.2),

$$P_N = F^{(N)}(T) G^{(N)}(Z), \tag{5.22}$$

and the probability that it is replaced at damage Z is, from (3.3),

$$P_Z = \sum_{j=0}^{N-1} F^{(j+1)}(T) \left[G^{(j)}(Z) - G^{(j+1)}(Z) \right], \tag{5.23}$$

that includes the probability that the total damage has exceeded Z at shock N. It is clearly seen that $P_T + P_N + P_Z = 1$.

Furthermore, the mean time to replacement is

$$T \sum_{j=0}^{N-1} \left[F^{(j)}(T) - F^{(j+1)}(T) \right] G^{(j)}(Z) + G^{(N)}(Z) \int_0^T t \, dF^{(N)}(t)$$

$$+ \sum_{j=0}^{N-1} \left[G^{(j)}(Z) - G^{(j+1)}(Z) \right] \int_0^T t \, dF^{(j+1)}(t)$$

$$= \sum_{j=0}^{N-1} G^{(j)}(Z) \int_0^T \left[F^{(j)}(t) - F^{(j+1)}(t) \right] dt, \tag{5.24}$$

that is equal to (3.5) by replacing $F_j(t)$ with $F^{(j)}(t)$. Similarly, the expected number of failures before replacement is

$$\sum_{j=1}^{N-1} \left[F^{(j)}(T) - F^{(j+1)}(T)\right] \sum_{i=1}^{j} \int_0^Z p(x)\, \mathrm{d}G^{(i)}(x) + F^{(N)}(T) \sum_{j=1}^{N-1} \int_0^Z p(x)\, \mathrm{d}G^{(j)}(x)$$

$$= \sum_{j=1}^{N-1} F^{(j)}(T) \int_0^Z p(x)\, \mathrm{d}G^{(j)}(x). \tag{5.25}$$

Let c_M be the cost of minimal repair, and c_k ($k = T, N, Z$) be the replacement cost at k. Then, the expected cost rate is, summing up $c_T P_T + c_N P_N + c_Z P_Z + c_M \times$ (5.25) and dividing by (5.24),

$$C(T, N, Z) = \frac{\begin{array}{l} c_Z - (c_Z - c_T) \sum_{j=0}^{N-1} \left[F^{(j)}(T) - F^{(j+1)}(T)\right] G^{(j)}(Z) \\ -(c_Z - c_N) F^{(N)}(T) G^{(N)}(Z) \\ +c_M \sum_{j=1}^{N-1} F^{(j)}(T) \int_0^Z p(x)\, \mathrm{d}G^{(j)}(x) \end{array}}{\sum_{j=0}^{N-1} G^{(j)}(Z) \int_0^T \left[F^{(j)}(t) - F^{(j+1)}(t)\right] \mathrm{d}t}. \tag{5.26}$$

5.4.2 Optimum Policies

We discuss analytically optimum T^*, N^*, and Z^* that minimize the expected cost rates when $p(x) = 1 - \mathrm{e}^{-\theta x}$ ($0 < 1/\theta < \infty$). In this case, the probability that the unit fails at shock j is

$$\int_0^\infty p(x)\, \mathrm{d}G^{(j)}(x) = \int_0^\infty (1 - \mathrm{e}^{-\theta x})\, \mathrm{d}G^{(j)}(x) = 1 - [G^*(\theta)]^j,$$

where $G^*(\theta)$ denotes the Laplace–Stieltjes transform of $G(x)$, i.e., $G^*(\theta) \equiv \int_0^\infty \mathrm{e}^{-\theta x} \mathrm{d}G(x) < 1$ for $\theta > 0$.

(1) Optimum T^*

A unit is replaced only at time T (Figure 5.2). Then, from (5.26),

$$C_1(T) \equiv \lim_{\substack{N \to \infty \\ Z \to \infty}} C(T, N, Z) = \frac{1}{T} \left[c_M \sum_{j=1}^{\infty} F^{(j)}(T) \left\{1 - [G^*(\theta)]^j\right\} + c_T \right], \tag{5.27}$$

that agrees with (5.2) of block replacement by replacing $\Phi(t)$ with $F(t)$ when $G^*(\theta) \equiv 0$. We seek an optimum time T^* that minimizes $C_1(T)$ when $G^*(\theta) > 0$. Differentiating $C_1(T)$ with respect to T and setting it equal to zero,

$$\sum_{j=1}^{\infty} \left[Tf^{(j)}(T) - F^{(j)}(T)\right] \left\{1 - [G^*(\theta)]^j\right\} = \frac{c_T}{c_M}, \tag{5.28}$$

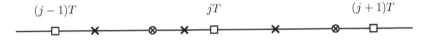

Fig. 5.2. Process for periodic replacement at time T

where $f(t)$ is a density function of $F(t)$ and $f^{(j)}(t)$ is the j-fold convolution of $f(t)$ with itself.

In particular, shocks occur in a Poisson process with rate λ, i.e., $F^{(j)}(t) = \sum_{i=j}^{\infty}[(\lambda t)^i/i!]e^{-\lambda t}$ $(j = 0, 1, 2, \cdots)$. Then, (5.28) is rewritten as

$$1 - \{1 + \lambda T\,[1 - G^*(\theta)]\}\,e^{-\lambda T[1-G^*(\theta)]} = \frac{1 - G^*(\theta)}{G^*(\theta)}\frac{c_T}{c_M}. \qquad (5.29)$$

The left-hand side of (5.29) is a gamma distribution of order 2 that increases from 0 to 1. Thus, we have the optimum policy:

(i) If $G^*(\theta)/[1 - G^*(\theta)] > c_T/c_M$, then there exist a finite and unique T^* that satisfies (5.29), and the resulting cost rate is

$$C_1(T^*) = \lambda c_M \left\{1 - G^*(\theta)e^{-\lambda T^*[1-G^*(\theta)]}\right\}. \qquad (5.30)$$

(ii) If $G^*(\theta)/[1 - G^*(\theta)] \le c_T/c_M$, then $T^* = \infty$, i.e., the unit is not be replaced, and $C_1(\infty) = \lambda c_M$.

It is of interest that

$$\sum_{j=1}^{\infty}\int_0^{\infty} e^{-\theta x}\,dG^{(j)}(x) = \int_0^{\infty} e^{-\theta x}\,dM_G(x) = \frac{G^*(\theta)}{1 - G^*(\theta)}$$

represents the expected number of nonfailures for an infinite interval, where $M_G(x) \equiv \sum_{j=1}^{\infty} G^{(j)}(x)$. In general, the expected number for actual models would be greater than the ratio c_T/c_M of two costs. Furthermore, from (5.29), T^* is given approximately by

$$\tilde{T} = \frac{1}{\lambda}\sqrt{\frac{1}{G^*(\theta)[1 - G^*(\theta)]}\frac{c_T}{c_M}},$$

and $T^* > \tilde{T}$.

(2) Optimum N^*

A unit is replaced only at shock N (Figure 5.3). Then, from (5.26),

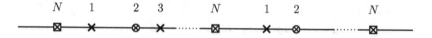

N   1   2   3   N   1   2   N

□ Planned replacement   ✗ Shock point   ⊗ Minimal repair at failure

Fig. 5.3. Process for replacement at shock N

$$C_2(N) \equiv \lim_{\substack{T \to \infty \\ Z \to \infty}} C(T, N, Z)$$

$$= \frac{\lambda}{N} \left[c_M \sum_{j=0}^{N-1} \{1 - [G^*(\theta)]^j\} + c_N \right] \quad (N = 1, 2, \cdots). \tag{5.31}$$

Forming the inequality $C_2(N+1) - C_2(N) \geq 0$,

$$\frac{1 - [G^*(\theta)]^N}{1 - G^*(\theta)} - N[G^*(\theta)]^N \geq \frac{c_N}{c_M} \quad (N = 1, 2, \cdots). \tag{5.32}$$

The left-hand side of (5.32) is strictly increasing to $1/[1 - G^*(\theta)]$. Thus, we have the optimum policy:

(i) If $1/[1 - G^*(\theta)] > c_N/c_M$, then there exists a finite and unique minimum number N^* $(1 \leq N^* < \infty)$ that satisfies (5.32), and its resulting cost rate is

$$\lambda c_M \left\{1 - [G^*(\theta)]^{N^*-1}\right\} < C_2(N^*) \leq \lambda c_M \left\{1 - [G^*(\theta)]^{N^*}\right\}. \tag{5.33}$$

(ii) If $1/[1 - G^*(\theta)] \leq c_N/c_M$, then $N^* = \infty$ and $C_2(\infty) = C_1(\infty)$.

It is clearly seen that if $1 - G^*(\theta) \geq c_N/c_M$, then $N^* = 1$.

It has been assumed until now that shocks occur in a renewal process. If shocks occur in a nonhomogeneous Poisson process with an intensity function $h(t)$ and a mean value function $H(t)$, as shown in (2.16), the mean time to the Nth shock is, from (1.29),

$$\sum_{j=0}^{N-1} \int_0^\infty \frac{[H(t)]^j}{j!} e^{-H(t)} \, dt,$$

and hence, the expected cost rate is

$$\tilde{C}_2(N) = \frac{c_M \sum_{j=0}^{N-1} \{1 - [G^*(\theta)]^j\} + c_N}{\sum_{j=0}^{N-1} \int_0^\infty p_j(t) \, dt} \quad (N = 1, 2, \cdots), \tag{5.34}$$

where $p_j(t) \equiv \{[H(t)]^j/j!\} e^{-H(t)}$ $(j = 0, 1, 2, \ldots)$. When $G^*(\theta) \equiv 0, \tilde{C}_2(N)$ agrees with (4.40).

We also seek an optimum N^* that minimizes $\widetilde{C}_2(N)$ in (5.34). Forming the inequality $\widetilde{C}_2(N+1) - \widetilde{C}_2(N) \geq 0$,

$$\{1 - [G^*(\theta)]^N\} \frac{\sum_{j=0}^{N-1} \int_0^\infty p_j(t)\,dt}{\int_0^\infty p_N(t)\,dt} - \sum_{j=0}^{N-1} \{1 - [G^*(\theta)]^j\} \geq \frac{c_N}{c_M}$$

$$(N = 1, 2, \ldots). \qquad (5.35)$$

It is assumed that the intensity function $h(t)$ is increasing . Then, letting $Q(N)$ be the left-hand side of (5.35), it can be proved that

$$Q(N+1) - Q(N) = \sum_{j=0}^{N} \int_0^\infty p_j(t)\,dt \left\{ \frac{1 - [G^*(\theta)]^{N+1}}{\int_0^\infty p_{N+1}(t)\,dt} - \frac{1 - [G^*(\theta)]^N}{\int_0^\infty p_N(t)\,dt} \right\} > 0,$$

because $\int_0^\infty p_N(t)dt$ is deceasing in N to $1/h(\infty)$ from (1.29). Thus, we have the optimum policy when $h(t)$ is increasing:

(i) If $Q(\infty) > c_N/c_M$, then there exists a finite and unique minimum number N^* $(1 \leq N^* < \infty)$ that satisfies (5.35), and its resulting cost rate is

$$\frac{c_M[1 - G^*(\theta)]^{N^*-1}}{\int_0^\infty p_{N^*-1}(t)\,dt} < \widetilde{C}_2(N^*) \leq \frac{c_M[1 - G^*(\theta)]^{N^*}}{\int_0^\infty p_{N^*}(t)\,dt}. \qquad (5.36)$$

(ii) If $Q(\infty) \leq c_N/c_M$, then $N^* = \infty$ and $\widetilde{C}_2(\infty) = c_M h(\infty)$.

Furthermore, we have the inequality

$$Q(N) \geq \frac{\{1 - [G^*(\theta)]^N\}}{\lambda \int_0^\infty p_N(t)\,dt},$$

where $1/\lambda \equiv \int_0^\infty e^{-H(t)}\,dt$, because $\int_0^\infty p_N(t)dt$ is deceasing in N. Therefore, if

$$\lim_{N \to \infty} \frac{\{1 - [G^*(\theta)]^N\}}{\lambda \int_0^\infty p_N(t)\,dt} = \frac{h(\infty)}{\lambda} > \frac{c_N}{c_M},$$

then a finite solution to (5.35) exists uniquely. Clearly, if $h(t)$ goes to ∞, as $t \to \infty$, then a finite N^* always exists.

(3) Optimum Z^*

A unit is replaced only at damage Z (Figure 5.4). Then, from (5.26),

$$C_3(Z) \equiv \lim_{\substack{T \to \infty \\ N \to \infty}} C(T, N, Z)$$

$$= \frac{c_M \int_0^Z p(x)\,dM_G(x) + c_Z}{[1 + M_G(Z)]/\lambda}. \qquad (5.37)$$

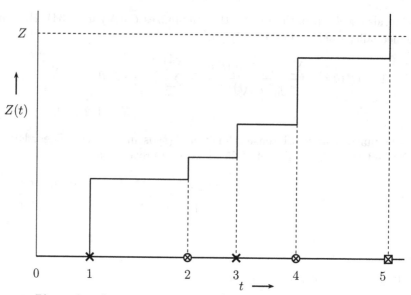

Fig. 5.4. Process for replacement at damage Z

Differentiating $C_3(Z)$ with respect to Z and setting it equal to zero,

$$\int_0^Z [1 + M_G(x)]\, dp(x) = \frac{c_Z}{c_M}, \tag{5.38}$$

that is strictly increasing in Z. Thus, if $\int_0^\infty [1 + M_G(x)] dp(x) > c_Z/c_M$, then there exists a finite and unique Z^* $(0 < Z^* < \infty)$ that satisfies (5.38).

In particular, when $p(x) = 1 - e^{-\theta x}$,

$$\int_0^\infty [1 + M_G(x)]\, dp(x) = \frac{1}{1 - G^*(\theta)}. \tag{5.39}$$

Therefore, we have the optimum policy:

(i) If $1/[1 - G^*(\theta)] > c_Z/c_M$, then there exists a finite and unique Z^* that satisfies (5.38), and its resulting cost rate is

$$C_3(Z^*) = \lambda c_M p(Z^*). \tag{5.40}$$

(ii) If $1/[1 - G^*(\theta)] \le c_Z/c_M$, then $Z^* = \infty$, and $C_3(\infty) = C_1(\infty)$.

Example 5.4. Table 5.3 presents the optimum time T^* satisfying (5.29) and expected cost rate $C_1(T^*)$ in (5.30), and the optimum number N^* satisfying (5.32) and expected cost rate $C_2(N^*)$ in (5.31) for c_k $(k = T, N)$ $= 5$–20

Table 5.3. Optimum time T^*, expected cost rate $C_1(T^*)/c_M$, and optimum shock number N^*, expected cost rate $C_2(N^*)/c_M$ when $c_M = 5$, $\lambda = 1$ and $G^*(\theta) = 0.9$

c_k	T^*	$C_1(T^*)/c_M$	N^*	$C_2(N^*)/c_M$
5	5.67	0.489	5	0.381
6	6.34	0.523	6	0.419
7	7.00	0.553	6	0.452
8	7.62	0.580	7	0.483
9	8.25	0.606	7	0.512
10	8.86	0.629	8	0.538
11	9.46	0.651	8	0.563
12	10.07	0.671	9	0.586
13	10.67	0.690	9	0.608
14	11.28	0.709	10	0.629
15	11.89	0.726	10	0.649
16	12.51	0.742	11	0.667
17	13.13	0.758	11	0.685
18	13.77	0.773	12	0.702
19	14.41	0.787	13	0.719
20	15.07	0.801	13	0.734

when $c_M = 5$, $\lambda = 1$, and $G^*(\theta) = 0.9$. In this case, finite T^* and N^* exist uniquely for $c_T < 45$ and $c_N < 50$, and the expected number of nonfailures is $G^*(\theta)/[1 - G^*(\theta)] = 9$.

If $c_N \le c_T$, then the replacement with shock N is better than that with time T, and if $c_N \ge c_T + c_M$, then the replacement with time T is better than that with shock N. In the case of $c_T < c_N < c_T + c_M$, for example, when $c_T = 10$ and $c_N = 14$, both replacement policies are almost the same. ∎

5.5 Modified Replacement Models

(1) Replacement with Threshold Level

Consider the periodic replacement policy in which a unit is replaced at time nT ($n = 1, 2, \dots$). If the total damage $Z(T)$ has exceeded a threshold level K between planned replacements, the total cost would be higher than anticipated [197]. The other assumptions are the same as those in Section 5.3 except minimal repair at failures. Let $c_0(x)$ be an additional replacement cost for the total damage x defined in (1) of Section 3.3. Then, the expected cost rate is, from (3.29),

$$C(T) = \frac{1}{T}\left[\sum_{j=0}^{\infty}[F^{(j)}(T) - F^{(j+1)}(T)]\left\{\int_0^K [c_T + c_0(x)]\,dG^{(j)}(x)\right.\right.$$

$$\left.\left. + \int_K^{\infty} [c_K + c_0(x)]\,dG^{(j)}(x)\right\}\right], \tag{5.41}$$

where c_T and c_K are the replacement cost at time nT when $Z(T) \le K$ and $Z(T) > K$, respectively.

(2) Replacement at the Next Shock over Time T

A unit is not replaced at time T. After T, it is replaced at the next shock and undergoes minimal repair at failures between replacements (see **(3)** of Section (3.3)). Because the mean time to replacement is, from (5.40) of [1],

$$\sum_{j=0}^{\infty} \int_0^T \left[\int_{T-t}^{\infty}(t+u)\,dF(u)\right]dF^{(j)}(t)$$

$$= T + \int_T^{\infty}\overline{F}(t)\,dt + \int_0^T\left[\int_{T-t}^{\infty}\overline{F}(u)\,du\right]dM_F(t),$$

the expected cost rate is, from (5.27),

$$\widetilde{C}_1(T) = \frac{c_M\sum_{j=1}^{\infty}F^{(j)}(T)\left\{1 - [G^*(\theta)]^j\right\} + c_T}{T + \int_T^{\infty}\overline{F}(t)\,dt + \int_0^T[\int_{T-t}^{\infty}\overline{F}(u)\,du]\,dM_F(t)}. \tag{5.42}$$

In particular, when $F(t) = 1 - e^{-\lambda t}$,

$$\frac{\widetilde{C}_1(T)}{\lambda} = \frac{c_M\left\{\lambda T - [G^*(\theta)/(1 - G^*(\theta))][1 - e^{-\lambda T[1-G^*(\theta)]}]\right\} + c_T}{\lambda T + 1}. \tag{5.43}$$

When $T = 0$, i.e., the unit is always replaced at the first shock, the expected cost rate is $\widetilde{C}_1(0) = \lambda c_T$, and when the unit is replaced never, it is $\widetilde{C}_1(\infty) = \lambda c_M$.

We seek an optimum time T^* $(0 \le T^* \le \infty)$ that minimizes $\widetilde{C}_1(T)$ in (5.43). Differentiating $\widetilde{C}_1(T)$ with respect to T and setting it equal to zero,

$$1 - (1 + \lambda T)G^*(\theta)e^{-\lambda T[1-G^*(\theta)]} + \frac{G^*(\theta)}{1 - G^*(\theta)}\left\{1 - e^{-\lambda T[1-G^*(\theta)]}\right\} = \frac{c_T}{c_M}. \tag{5.44}$$

The left-hand side of (5.44) is strictly increasing from $1 - G^*(\theta)$ to $1/[1 - G^*(\theta)]$. Thus, we have the optimum policy:

(i) If $1 - G^*(\theta) \ge c_T/c_M$, then $T^* = 0$.

(ii) If $1 - G^*(\theta) < c_T/c_M < 1/[1 - G^*(\theta)]$, then there exists a finite and unique $T^* (0 < T^* < \infty)$ that satisfies (5.44), and the resulting cost rate is

$$\tilde{C}_1(T^*) = \lambda c_M \left\{ 1 - G^*(\theta)e^{-\lambda T^*[1 - G^*(\theta)]} \right\}. \tag{5.45}$$

(iii) If $1/[1 - G^*(\theta)] \leq c_T/c_M$, then $T^* = \infty$.

For example, when $G^*(\theta) = 0.9$, $T^* = 0$ for $c_T/c_M \leq 0.1$, $0 < T^* < \infty$ for $0.1 < c_T/c_M < 10$, and $T^* = \infty$ for $c_T/c_M \geq 10$. It is clearly seen that T^* to satisfy (5.44) is smaller than that to satisfy (5.29).

(3) Replacement at the Next Shock over Damage Z

A unit is replaced at the next shock when the total damage has exceeded a threshold level Z. Then, the expected number of failures before replacement is

$$\sum_{j=0}^{\infty} \left\{ \int_0^Z p(x)\,dG^{(j)}(x) + \int_0^Z \left[\int_{Z-x}^{\infty} p(x+y)\,dG(y) \right] dG^{(j)}(x) \right\}$$

$$= \int_0^{\infty} p(x)\,dG(x) + \int_0^Z \left[\int_0^{\infty} p(x+y)\,dG(y) \right] dM_G(x).$$

Furthermore, the mean time to replacement increases by the mean shock time $1/\lambda$ in the denominator of (5.37). Thus, the expected cost rate is

$$\frac{\tilde{C}_3(Z)}{\lambda} = \frac{c_M \left\{ \int_0^{\infty} p(x)\,dG(x) + \int_0^Z \left[\int_0^{\infty} p(x+y)\,dG(y) \right] dM_G(x) \right\} + c_Z}{2 + M_G(Z)}. \tag{5.46}$$

Differentiating $\tilde{C}_3(Z)$ with respect to Z and setting it equal to zero,

$$[2 + M_G(Z)] \int_0^{\infty} p(Z+x)\,dG(x) - \int_0^Z \left[\int_0^{\infty} p(x+y)\,dG(y) \right] dM_G(x)$$

$$- \int_0^{\infty} p(x)\,dG(x) = \frac{c_Z}{c_M}. \tag{5.47}$$

Letting $Q(Z)$ be the left-hand side of (5.47), we easily see that $Q(Z)$ is strictly increasing from $\int_0^{\infty} p(x)dG(x)$ to $Q(\infty)$. Thus, we have the optimum policy:

(i) If $\int_0^{\infty} p(x)dG(x) \geq c_Z/c_M$, then $Z^* = 0$, and

$$\frac{\tilde{C}_3(0)}{\lambda} = \frac{c_M \int_0^{\infty} p(x)\,dx + c_Z}{2}.$$

(ii) If $\int_0^{\infty} p(x)dG(x) < c_Z/c_M < Q(\infty)$, then there exists a finite and unique $Z^* (0 < Z^* < \infty)$ that satisfies (5.47), and the resulting cost rate is

$$\tilde{C}_3(Z^*) = \lambda c_M \int_0^{\infty} p(Z^* + x)\,dG(x). \tag{5.48}$$

(iii) If $Q(\infty) \leq c_Z/c_M$, then $Z^* = \infty$.

It is clearly seen that

$$Q(Z) \geq 2 \int_0^\infty p(Z + x)\, \mathrm{d}G(x) - \int_0^\infty p(x)\, \mathrm{d}G(x),$$

because $p(x)$ is increasing in x. Therefore, if $2 - \int_0^\infty p(x)\mathrm{d}G(x) > c_Z/c_M$, i.e., $\int_0^\infty [1 - p(x)]\mathrm{d}G(x) > (c_Z - c_M)/c_M$ then a finite Z^* exists.

Example 5.5.   Suppose that $G(x) = 1 - \mathrm{e}^{-\mu x}$ and $p(x) = 1 - \mathrm{e}^{-\theta x}$. Then, we compare the expected cost rates $C_3(Z)$ in (5.37) and $\widetilde{C}_3(Z)$ in (5.46) numerically. Under such assumptions, the expected cost rate $C_3(Z)$ is rewritten as

$$\frac{C_3(Z)}{\lambda} = \frac{c_M[\mu Z - (\mu/\theta)(1 - \mathrm{e}^{-\theta Z})] + c_Z}{1 + \mu Z},$$

and if $(\mu+\theta)/\theta > c_Z/c_M$, then there exists a finite and unique Z_1^* that satisfies

$$\left(1 + \frac{\mu}{\theta}\right)\left(1 - \mathrm{e}^{-\theta Z}\right) - \mu Z \mathrm{e}^{-\theta Z} = \frac{c_Z}{c_M}.$$

The expected cost rate $\widetilde{C}_3(Z)$ is

$$\frac{\widetilde{C}_3(Z)}{\lambda} = \frac{c_M[\theta/(\mu + \theta)] + \mu\left\{Z - [\mu/(\theta(\mu + \theta))](1 - \mathrm{e}^{-\theta Z})\right\} + c_Z}{2 + \mu Z},$$

and if $(\mu + \theta)/\theta > c_Z/c_M > \theta/(\mu + \theta)$, then there exists a finite and unique Z_2^* that satisfies

$$1 - \frac{\mu}{\mu + \theta}(1 + \mu Z)\mathrm{e}^{-\theta Z} + \frac{\mu}{\theta}(1 - \mathrm{e}^{-\theta Z}) = \frac{c_Z}{c_M}.$$

Because

$$1 - \frac{\mu}{\mu + \theta}(1 + \mu Z)\mathrm{e}^{-\theta Z} + \frac{\mu}{\theta}(1 - \mathrm{e}^{-\theta Z}) > \left(1 + \frac{\mu}{\theta}\right)\left(1 - \mathrm{e}^{-\theta Z}\right) - \mu Z \mathrm{e}^{-\theta Z},$$

$Z_1^* > Z_2^*$.

Table 5.4 presents the optimum values of Z_1^* and Z_2^* that minimize $C_3(Z)$ and $\widetilde{C}_3(Z)$, respectively, and their resulting cost rates $C_3(Z_1^*)/\lambda$ and $\widetilde{C}_3(Z_2^*)/\lambda$ for $c_Z = 5\text{--}20$ when $c_M = 5$ and $G^*(\theta) = 0.9$, i.e., $\mu/\theta = 9$. In this case, both finite and positive Z_1^* and Z_2^* exist uniquely for $0.5 < c_Z < 50$, and $C_3(Z_1^*) < \widetilde{C}_3(Z_2^*)$ and $\theta Z_1^* < \theta/\mu + \theta Z_2^*$. However, their differences between two expected costs become smaller, as c_Z becomes larger. If the replacement cost c_Z is less than that of **(3)** in Section 5.4.2, this policy might be more useful than the policy of **(3)**. ∎

Table 5.4. Optimum damage level Z_1^*, expected cost rate $C_3(Z_1^*)$, and damage level Z_2^*, expected cost rate $\widetilde{C}_3(Z_2^*)$ when $c_M = 5$ and $\mu/\theta = 9$

c_Z	θZ_1^*	$C_3(Z_1^*)/\lambda$	θZ_2^*	$\widetilde{C}_3(Z_2^*)/\lambda$
5	0.437	1.770	0.342	1.804
6	0.498	1.963	0.402	1.991
7	0.557	2.136	0.461	2.161
8	0.615	2.296	0.517	2.317
9	0.670	2.443	0.573	2.462
10	0.726	2.581	0.627	2.597
11	0.781	2.709	0.682	2.724
12	0.835	2.830	0.735	2.843
13	0.889	2.944	0.789	2.956
14	0.943	3.053	0.843	3.063
15	0.997	3.155	0.897	3.165
16	1.052	3.253	0.951	3.262
17	1.107	3.347	1.006	3.354
18	1.162	3.435	1.061	3.443
19	1.218	3.521	1.117	3.528
20	1.275	3.603	1.174	3.609

6

Preventive Maintenance Policies

Most operating units are repaired or replaced when they have failed. If a failed unit undergoes repair, it begins to operate again after the repair completion. However, it may require much time and high cost to repair a failed unit. It may sometimes be necessary to maintain a unit to prevent failures. Some maintenance after failure and before failure is called *corrective maintenance* (CM) and *preventive maintenance* (PM), respectively. Optimum PM policies for some units were summarized [1, 200–202]. The modified PM policy that is planned only at periodic times was proposed in Section 6.3 of [1].

PM actions are generally grouped into time maintenance that is based on the planned time, age, or usage time of a unit, and monitored maintenance or condition-based maintenance that is based on the condition of a unit [203]. The first maintenance corresponds to the replacement policies discussed in Chapters 3–5 in [1] and the maintenance that is done at a planned time T or number N in Chapters 3–5. The latter maintenance is done by monitoring one or more variables charactering the wear, fatigue, and damage of an operating unit and corresponds to the maintenance that is done at a damage level Z or at a shock number N in Chapters 3–5.

This chapter takes up the modified PM policy [56] and applies it to a condition-based PM of a cumulative damage model, where the CM is done immediately when the total damage due to shocks has exceeded a failure level K. The test to investigate some characteristics of an operating unit is planned at periodic times nT $(n = 1, 2, \cdots)$. We can know the characteristics such as the damage and the shock number only through tests, and if necessary, we do some appropriate maintenance.

In Section 6.1, if the total damage has exceeded a threshold level Z $(0 \leq Z \leq K)$, the PM is done at the first planned time, when shocks occur in a nonhomogeneous Poisson process. The expected cost rate is obtained, and an optimum Z^* that minimizes it is discussed analytically. Furthermore, in Section 6.2, the modified PM models, where (1) the failure is detected only through tests, (2) the PM is done when the total number of shocks has exceeded a threshold number N, and (3) the PM is done at time NT, are proposed. The expected cost rates of each model are obtained, and a numerical example to compare them is given.

6.1 Condition-based Preventive Maintenance

We consider a condition-based PM policy where the condition of an operating unit is monitored at inspection times. If the condition is normal, the operation is continued. However, if the condition reaches a previously determined threshold level of resistance to failure, the PM is done before failure. Such PM policies have been actually in use for engines, mainflames, control systems of aircraft [204], and plants in the chemical and machine industries.

Condition-based maintenance models for a deteriorating system are generally classified into continuous wear processes [192,193] and Markovian deterioration processes [205–208]. In the former case, the preventive replacement level of a one-unit system whose condition is monitored at inspection times was considered, and optimum levels to minimize the expected cost and the availability were derived [194–196, 209–211]. This was extended to a two-unit series system [196].

This section adopts the condition-based PM policy for a cumulative damage model: A unit suffers damage due to shocks, and fails when the total amount of additive damage has exceeded a failure level K. Then, the CM is done immediately. The test is planned at periodic times nT $(n = 1, 2, \cdots)$ to prevent failures, where T (> 0) means a week, a month, or a year. It is assumed that we can know the total damage to a unit only through tests. If the total damage has exceeded a threshold level Z $(0 \leq Z \leq K)$ at time nT, the PM or overhaul is done before failure. Otherwise, no PM should be done.

Suppose that shocks occur in a nonhomogeneous Poisson process. Then, using the theory of a Poisson process and the results of Section 6.3 of [1], we obtain the expected cost rate and determine an optimum damage level Z^* that minimizes it. In particular, when shocks occur in a Poisson process, an optimum Z^* is given by a unique solution of the equation.

6.1.1 Expected Cost Rate

Consider a unit that should operate over an infinite time interval: Shocks occur in a nonhomogeneous Poisson process with an intensity function $h(t)$ and a mean value function $H(t)$, i.e., $H(t) \equiv \int_0^t h(u)\mathrm{d}u$ represents the expected number of shocks in $[0, t]$, and $p_j[H(t)] \equiv \{[H(t)]^j / j!\}\, \mathrm{e}^{-H(t)}$ $(j = 0, 1, 2, \cdots)$ is the probability that j shocks occur exactly in $[0, t]$. In addition, random variables $\{W_j\}$ $(j = 1, 2, \cdots)$ denote an amount of damage due to the jth shock and are nonnegative, independent, and identically distributed. Each W_j is statistically estimated and has an identical distribution $G(x) \equiv \Pr\{W_j \leq x\}$ $(j = 1, 2, \cdots)$. Each amount of damage is additive, and $G^{(j)}(x)$ denotes the

j-fold Stieltjes convolution of $G(x)$ with itself $(j = 1, 2, \cdots)$ and $G^{(0)}(x) \equiv 1$ for $x \geq 0$. A unit fails only when the total damage has exceeded a failure level K, and then the CM is done.

Under the above assumptions, the test is planned at periodic times nT $(n = 1, 2, \cdots)$ to investigate the total damage, where a positive T is given. If the total damage has exceeded a threshold level Z $(0 \leq Z \leq K)$ during $(nT, (n+1)T]$ $(n = 0, 1, 2, \cdots)$, then its damage can be known through the test at time $(n + 1)T$, and the PM is done immediately (Figure 6.1). Otherwise, the unit is left as it is. The unit becomes as good as a new one at each PM or CM, $i.e.$, the PM is perfect. The imperfect PM policy for a cumulative damage model will be discussed in Chapter 7. The times required for any test and maintenance are negligible, $i.e.$, the time considered here is measured only by the total operating time.

We obtain the expected cost rate by a method similar to Section 6.3 of [1] and [56]. The probability that j shocks occur during $[0, nT]$ and the total damage is less than Z, and i shocks occur during $(nT, (n+1)T]$ and the total damage has exceeded K, is

$$p_j[H(nT)]\, p_i[H((n + 1)T) - H(nT)]$$
$$\times \Pr\left\{W_1 + \cdots + W_j \leq Z \text{ and } W_1 + \cdots + W_j + \cdots + W_{j+i} > K\right\}$$
$$= p_j[H(nT)]p_i[H((n + 1)T) - H(nT)] \int_0^Z [1 - G^{(i)}(K - x)]\, dG^{(j)}(x).$$

Thus, the probability that the unit fails and the CM is done immediately is

$$\sum_{n=0}^{\infty} \sum_{j=0}^{\infty} p_j[H(nT)] \sum_{i=0}^{\infty} p_i[H((n + 1)T) - H(nT)]$$
$$\times \int_0^Z [1 - G^{(i)}(K - x)]\, dG^{(j)}(x). \tag{6.1}$$

Conversely, the probability that the PM is done at time $(n + 1)T$ $(n = 0, 1, 2, \cdots)$ when the total damage is between Z and K during $(nT, (n+1)T]$ is

$$\sum_{n=0}^{\infty} \sum_{j=0}^{\infty} p_j[H(nT)] \sum_{i=0}^{\infty} p_i[H((n + 1)T) - H(nT)]$$
$$\times \Pr\left\{W_1 + \cdots + W_j \leq Z \text{ and } Z < W_1 + \cdots + W_j + \cdots + W_{j+i} \leq K\right\}$$
$$= \sum_{n=0}^{\infty} \sum_{j=0}^{\infty} p_j[H(nT)] \sum_{i=0}^{\infty} p_i[H((n + 1)T) - H(nT)]$$
$$\times \int_0^Z [G^{(i)}(K - x) - G^{(i)}(Z - x)]\, dG^{(j)}(x). \tag{6.2}$$

It is proved that $(6.1) + (6.2) = 1$, because, from the reproductive property of a Poisson distribution,

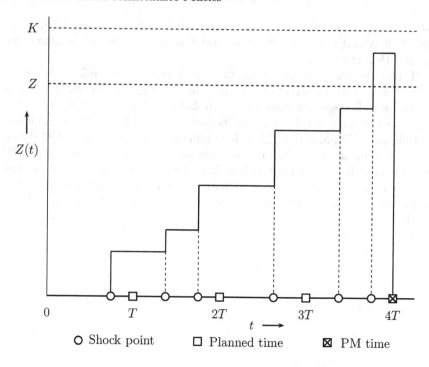

Fig. 6.1. Process for PM at damage Z

$$\sum_{n=0}^{\infty}\sum_{j=0}^{\infty}p_j[H(nT)]\sum_{i=0}^{\infty}p_i[H((n+1)T)-H(nT)][G^{(j)}(Z)-G^{(i+j)}(Z)]$$

$$=\sum_{n=0}^{\infty}\left\{\sum_{j=0}^{\infty}p_j[H(nT)]G^{(j)}(Z)\right.$$

$$\left.-\sum_{i=0}^{\infty}G^{(i)}(Z)\sum_{j=0}^{i}p_j[H(nT)]p_{i-j}[H((n+1)T)-H(nT)]\right\}$$

$$=\sum_{n=0}^{\infty}\left\{\sum_{j=0}^{\infty}p_j[H(nT)]G^{(j)}(Z)-\sum_{i=0}^{\infty}p_i[H((n+1)T)]G^{(i)}(Z)\right\}=1.$$

The mean time to either PM or CM is

$$\sum_{n=0}^{\infty} [(n+1)T] \sum_{j=0}^{\infty} p_j [H(nT)] \sum_{i=0}^{\infty} p_i [H((n+1)T) - H(nT)]$$

$$\times \int_0^Z [G^{(i)}(K-x) - G^{(i)}(Z-x)] \, \mathrm{d}G^{(j)}(x)$$

$$+ \sum_{n=0}^{\infty} \sum_{j=0}^{\infty} p_j [H(nT)] \sum_{i=0}^{\infty} \int_{nT}^{(n+1)T} t p_i [H(t) - H(nT)] h(t) \, \mathrm{d}t$$

$$\times \int_0^Z [G^{(i)}(K-x) - G^{(i+1)}(K-x)] \, \mathrm{d}G^{(j)}(x)$$

$$= \sum_{n=0}^{\infty} \sum_{j=0}^{\infty} p_j [H(nT)] \sum_{i=0}^{\infty} \int_0^Z G^{(i)}(K-x) \, \mathrm{d}G^{(j)}(x)$$

$$\times \int_{nT}^{(n+1)T} p_i [H(t) - H(nT)] \, \mathrm{d}t. \tag{6.3}$$

Let c_Z be the PM cost before failure and c_K be the CM cost after failure with $c_K > c_Z$. Then, the expected cost rate is, summing up $c_K \times (6.1) + c_Z \times (6.2)$ and dividing by (6.3),

$$C_1(Z) = \frac{c_Z + (c_K - c_Z) \sum_{n=0}^{\infty} \sum_{j=0}^{\infty} p_j [H(nT)]}{\sum_{n=0}^{\infty} \sum_{j=0}^{\infty} p_j [H(nT)] \sum_{i=0}^{\infty} \int_0^Z G^{(i)}(K-x) \, \mathrm{d}G^{(j)}(x)} \cdot \frac{\times \sum_{i=0}^{\infty} p_i [H((n+1)T) - H(nT)] \int_0^Z [1 - G^{(i)}(K-x)] \, \mathrm{d}G^{(j)}(x)}{\times \int_{nT}^{(n+1)T} p_i [H(t) - H(nT)] \, \mathrm{d}t}$$

$$\tag{6.4}$$

Each amount of damage during $(nT, (n+1)T]$ is investigated only through tests and has an identical distribution $G(x)$ for all n ($n = 0, 1, 2, \cdots$). This corresponds to a cumulative damage model where shocks occur at every constant time T and the total damage is known at the end of each period. In this case, the expected cost rate is obtained by replacing $1/\lambda$ with T in (3.24), and the optimum policy has been derived in (3) of Section 3.2.

Next, a failure level K is statistically distributed, *i.e.*, K is a random variable and has a general distribution $L(x) \equiv \Pr\{K \le x\}$. Then, the expected cost rate in (6.4) is rewritten as

$$C_1(Z) = \frac{c_Z + (c_K - c_Z) \sum_{n=0}^{\infty} \sum_{j=0}^{\infty} p_j [H(nT)] \sum_{i=0}^{\infty} p_i [H((n+1)T) - H(nT)]}{\sum_{n=0}^{\infty} \sum_{j=0}^{\infty} p_j [H(nT)] \sum_{i=0}^{\infty} \int_0^Z \{\int_0^{\infty} [1 - L(x+y)] \, \mathrm{d}G^{(i)}(y)\} \, \mathrm{d}G^{(j)}(x)} \cdot \frac{\times \int_0^Z \{\int_0^{\infty} [L(x+y) - L(x)] \, \mathrm{d}G^{(i)}(y)\} \, \mathrm{d}G^{(j)}(x)}{\times \int_{nT}^{(n+1)T} p_i [H(t) - H(nT)] \, \mathrm{d}t}$$

$$\tag{6.5}$$

6.1.2 Optimum Policy

We seek an optimum threshold level Z^* that minimizes the expected cost rate $C_1(Z)$ in (6.4) when shocks occur in a Poisson process, $i.e.$, $p_j[H(nT)] = [(n\lambda T)^j/j!]e^{-n\lambda T} \equiv p_j(n\lambda T)$ $(j = 0, 1, 2, \cdots)$. Differentiating $C_1(Z)$ with respect to Z and setting it equal to zero,

$$Q_1(Z) \sum_{n=0}^{\infty} \sum_{j=0}^{\infty} p_j(n\lambda T) \sum_{i=0}^{\infty} \int_0^T p_i(\lambda t)\,dt \int_0^Z G^{(i)}(K-x)\,dG^{(j)}(x)$$

$$- \sum_{n=0}^{\infty} \sum_{j=0}^{\infty} p_j(n\lambda T) \sum_{i=0}^{\infty} p_i(\lambda T) \int_0^Z [1 - G^{(i)}(K-x)]\,dG^{(j)}(x) = \frac{c_Z}{c_K - c_Z},$$

$$(6.6)$$

where

$$Q_1(Z) \equiv \frac{\sum_{i=0}^{\infty} p_i(\lambda T)[1 - G^{(i)}(K-Z)]}{\sum_{i=0}^{\infty} \int_0^T p_i(\lambda t)\,dt\, G^{(i)}(K-Z)}.$$

It can be easily seen that $Q_1(Z)$ is increasing in Z from $Q_1(0)$ to λ. Denoting the left-hand side of (6.6) by $Q_2(Z)$, $Q_2(0) = 0$,

$$Q_2(K) = \sum_{n=0}^{\infty} \sum_{i=0}^{\infty} G^{(i)}(K) \int_{nT}^{(n+1)T} \lambda p_i(\lambda t)\,dt - 1 = \sum_{i=1}^{\infty} G^{(i)}(K),$$

$$\frac{dQ_2(Z)}{dZ} = \frac{dQ_1(Z)}{dZ} \sum_{n=0}^{\infty} \sum_{j=0}^{\infty} p_j(n\lambda T) \sum_{i=0}^{\infty} \int_0^T p_i(\lambda t)\,dt \int_0^Z G^{(i)}(K-x)\,dG^{(j)}(x).$$

It is assumed that the distribution $G(x)$ of each amount of damage due to shocks is continuous and strictly increasing. Then, $Q_2(Z)$ is also strictly increasing from 0 to $M_G(K) \equiv \sum_{j=1}^{\infty} G^{(j)}(K)$ that represents the expected number of shocks before the failure. Therefore, we have the following optimum policy:

(i) If $M_G(K) > c_Z/(c_K - c_Z)$, then there exists a unique Z^* $(0 < Z^* < K)$ that satisfies (6.6), and the resulting cost rate is

$$C_1(Z^*) = (c_K - c_Z)Q_1(Z^*). \qquad (6.7)$$

(ii) If $M_G(K) \le c_Z/(c_K - c_Z)$, then $Z^* = K$, and the CM is done after failure. In this case, the expected cost rate is

$$\frac{C_1(K)}{\lambda} = \frac{c_K}{1 + M_G(K)}, \qquad (6.8)$$

that agrees with (3.12).

This policy will be applied to a garbage collection model in Section 8.3, and an optimum level Z^* is computed numerically in Example 8.3.

6.2 Modified Models

We show the following modified models: (1) any failures are detected only through tests, (2) the PM is done when the total number of shocks has exceeded a threshold number N, and (3) the PM is done at time NT. The expected cost rates of each model are obtained.

(1) PM only at Test

Suppose that any failures are detected only through tests. When the unit fails during $(nT, (n+1)T]$, it is not detected immediately, but is detected only at time $(n+1)T$ and the CM is done. Then, the mean time to either PM or CM is

$$\sum_{n=0}^{\infty}[(n+1)T]\sum_{j=0}^{\infty}p_j[H(nT)]\sum_{i=0}^{\infty}p_i[H((n+1)T)-H(nT)]$$

$$\times\left\{\int_0^Z[1-G^{(i)}(K-x)]\,\mathrm{d}G^{(j)}(x)\right.$$

$$\left.+\int_0^Z[G^{(i)}(K-x)-G^{(i)}(Z-x)]\,\mathrm{d}G^{(j)}(x)\right\}$$

$$=T\sum_{n=0}^{\infty}\sum_{j=0}^{\infty}p_j[H(nT)]G^{(j)}(Z).\tag{6.9}$$

Furthermore, the mean time from a failure to its detection is, from (6.3),

$$\sum_{n=0}^{\infty}\sum_{j=0}^{\infty}p_j[H(nT)]\sum_{i=0}^{\infty}\int_{nT}^{(n+1)T}[(n+1)T-t]p_i[H(t)-H(nT)]h(t)\,\mathrm{d}t$$

$$\times\int_0^Z[G^{(i)}(K-x)-G^{(i+1)}(K-x)]\,\mathrm{d}G^{(j)}(x)$$

$$=\sum_{n=0}^{\infty}\sum_{j=0}^{\infty}p_j[H(nT)]\sum_{i=0}^{\infty}\int_0^Z[1-G^{(i)}(K-x)]\,\mathrm{d}G^{(j)}(x)$$

$$\times\int_{nT}^{(n+1)T}p_i[H(t)-H(nT)]\,\mathrm{d}t,\tag{6.10}$$

where note that (6.3) + (6.10) = (6.9). From this relation,

$$T\sum_{n=0}^{\infty}\sum_{j=0}^{\infty}p_j[H(nT)]G^{(j)}(K)\geq\sum_{j=0}^{\infty}G^{(j)}(K)\int_0^{\infty}p_j[H(t)]\,\mathrm{d}t,\tag{6.11}$$

that is the mean time to failure given in (2.19).

Let c_D be the loss cost per unit of time elapsed between a failure and its detection. Then, the expected cost rate is, from (6.4),

$$
\widetilde{C}_1(Z) = \frac{
\begin{aligned}
& c_Z + (c_K - c_Z) \sum_{n=0}^{\infty} \sum_{j=0}^{\infty} p_j [H(nT)] \\
& \times \sum_{i=0}^{\infty} p_i [H((n+1)T) - H(nT)] \int_0^Z [1 - G^{(i)}(K - x)] \, dG^{(j)}(x) \\
& - c_D \sum_{n=0}^{\infty} \sum_{j=0}^{\infty} p_j [H(nT)] \sum_{i=0}^{\infty} \int_0^Z G^{(i)}(K - x) \, dG^{(j)}(x) \\
& \times \int_{nT}^{(n+1)T} p_i [H(t) - H(nT)] \, dt
\end{aligned}
}{
T \sum_{n=0}^{\infty} \sum_{j=0}^{\infty} p_j [H(nT)] G^{(j)}(Z)
}
$$

$$
+ c_D. \tag{6.12}
$$

Compared with the expected cost rate $C_1(Z)$ in (6.4), $\widetilde{C}_1(Z)$ is smaller than $C_1(Z)$ when $c_D = 0$, and is larger as c_D increases. Thus, if the PM and CM costs are the same, $\widetilde{C}_1(Z)$ would be larger than $C_1(Z)$ when c_D is greater than some fixed cost.

When shocks occur in a Poisson process with rate λ, the expected cost rate $\widetilde{C}_1(Z)$ is rewritten as

$$
\widetilde{C}_1(Z) = \frac{
\begin{aligned}
& c_Z + (c_K - c_Z) \sum_{n=0}^{\infty} \sum_{j=0}^{\infty} p_j(n\lambda T) \sum_{i=0}^{\infty} p_i(\lambda T) \\
& \times \int_0^Z [1 - G^{(i)}(K - x)] \, dG^{(j)}(x) \\
& - c_D \sum_{n=0}^{\infty} \sum_{j=0}^{\infty} p_j(n\lambda T) \sum_{i=0}^{\infty} \int_0^T p_i(\lambda t) \, dt \\
& \times \int_0^Z G^{(i)}(K - x) \, dG^{(j)}(x)
\end{aligned}
}{
T \sum_{n=0}^{\infty} \sum_{j=0}^{\infty} p_j(n\lambda T) G^{(j)}(Z)
} + c_D. \tag{6.13}
$$

To find an optimum Z^* that minimizes $\widetilde{C}_1(Z)$, differentiating $\widetilde{C}_1(Z)$ with respect to Z and setting it equal to zero,

$$
\sum_{n=0}^{\infty} \sum_{j=0}^{\infty} p_j(n\lambda T) \sum_{i=0}^{\infty} \left[p_i(\lambda T) + \frac{c_D}{c_K - c_Z} \int_0^T p_i(\lambda t) \, dt \right]
$$

$$
\times \int_{K-Z}^K G^{(j)}(K - x) \, dG^{(i)}(x) = \frac{c_Z}{c_K - c_Z}. \tag{6.14}
$$

Denoting the left-hand side of (6.14) by $\widetilde{Q}(Z)$, we easily find that $\widetilde{Q}(Z)$ is strictly increasing from 0 to

$$
\widetilde{Q}(K) = \sum_{n=0}^{\infty} \sum_{j=0}^{\infty} p_j(n\lambda T) G^{(j)}(K) + \frac{c_D/\lambda}{c_K - c_Z} \sum_{j=0}^{\infty} G^{(j)}(K) - 1.
$$

Therefore, we have the following optimum policy:

(i) If $\widetilde{Q}(K) > c_Z/(c_K - c_Z)$, then there exists a unique Z^* $(0 < Z^* < K)$ that satisfies (6.14), and the resulting cost rate is

$$
\widetilde{C}_1(Z^*) = \frac{1}{T} \sum_{i=0}^{\infty} \left[1 - G^{(i)}(K - Z^*) \right] \left[(c_K - c_Z) p_i(\lambda T) + c_D \int_0^T p_i(\lambda t) \, dt \right].
$$

$$
\tag{6.15}
$$

(ii) If $\tilde{Q}(K) \le c_Z/(c_K - c_Z)$, then $Z^* = K$, and the expected cost rate is

$$\tilde{C}_1(K) = \frac{c_K - (c_D/\lambda)\sum_{j=0}^{\infty} G^{(j)}(K)}{T\sum_{n=0}^{\infty}\sum_{j=0}^{\infty} p_j(n\lambda T)G^{(j)}(K)} + c_D. \qquad (6.16)$$

From (6.11), because we have the inequality

$$\tilde{Q}(K) \ge \left(\frac{1}{\lambda T} + \frac{c_D/\lambda}{c_K - c_Z}\right)[1 + M_G(K)] - 1,$$

if

$$\frac{1 + M_G(K)}{\lambda T} > \frac{c_K}{c_K - c_Z + T c_D},$$

then a unique Z^* to satisfy (6.14) exists.

(2) PM at Shock Number

Suppose that the number of shocks is known only through tests. When the total number of shocks has exceeded a prespecified number N before failure during $(nT, (n+1)T]$, the PM is done at time $(n+1)T$. Then, by a method similar to (6.1) and (6.2), the probability that the CM is done after failure is

$$\sum_{n=0}^{\infty}\sum_{j=0}^{N-1} p_j[H(nT)]\sum_{i=0}^{\infty} p_i[H((n+1)T) - H(nT)][G^{(j)}(K) - G^{(i+j)}(K)], \quad (6.17)$$

and the probability that the PM is done before failure is

$$\sum_{n=0}^{\infty}\sum_{j=0}^{N-1} p_j[H(nT)]\sum_{i=N-j}^{\infty} p_i[H((n+1)T) - H(nT)]G^{(i+j)}(K), \qquad (6.18)$$

where note that $(6.17) + (6.18) = 1$. The mean time to either PM or CM is

$$\sum_{n=0}^{\infty}[(n+1)T]\sum_{j=0}^{N-1} p_j[H(nT)]\sum_{i=N-j}^{\infty} p_i[H((n+1)T) - H(nT)]G^{(i+j)}(K)$$

$$+ \sum_{n=0}^{\infty}\sum_{j=0}^{N-1} p_j[H(nT)]\sum_{i=0}^{\infty}[G^{(i+j)}(K) - G^{(i+j+1)}(K)]$$

$$\times \int_{nT}^{(n+1)T} t\, p_i[H(t) - H(nT)]h(t)\,dt$$

$$= \sum_{n=0}^{\infty}\sum_{j=0}^{N-1} p_j[H(nT)]\sum_{i=0}^{\infty} G^{(i+j)}(K)\int_{nT}^{(n+1)T} p_i[H(t) - H(nT)]\,dt. \quad (6.19)$$

Therefore, the expected cost rate is, summing up $c_K \times (6.17) + c_N \times (6.18)$ and dividing by (6.19),

$$C_2(N) = \frac{\begin{aligned} &c_N + (c_K - c_N)\sum_{n=0}^{\infty}\sum_{j=0}^{N-1} p_j[H(nT)] \\ &\times \sum_{i=0}^{\infty} p_i[H((n+1)T) - H(nT)][G^{(j)}(K) - G^{(i+j)}(K)] \end{aligned}}{\begin{aligned} &\sum_{n=0}^{\infty}\sum_{j=0}^{N-1} p_j[H(nT)]\sum_{i=0}^{\infty} G^{(i+j)}(K) \\ &\times \int_{nT}^{(n+1)T} p_i[H(t) - H(nT)]\,dt \end{aligned}}$$

$$(N = 1, 2, \cdots), \qquad (6.20)$$

where c_N is the PM cost at shock N.

If the failure is detected only through tests in the same way as (1), then the mean time to either PM or CM is

$$\sum_{n=0}^{\infty}[(n+1)T]\sum_{j=0}^{N-1} p_j[H(nT)]$$

$$\times \left\{ \sum_{i=0}^{\infty} p_i[H((n+1)T) - H(nT)][G^{(j)}(K) - G^{(i+j)}(K)] \right.$$

$$\left. + \sum_{i=N-j}^{\infty} p_i[H((n+1)T) - H(nT)]G^{(i+j)}(K) \right\}$$

$$= T\sum_{n=0}^{\infty}\sum_{j=0}^{N-1} p_j[H(nT)]G^{(j)}(K), \qquad (6.21)$$

and the mean time from a failure to its detection is

$$\sum_{n=0}^{\infty}\sum_{j=0}^{N-1} p_j[H(nT)]\sum_{i=0}^{\infty}[G^{(i+j)}(K) - G^{(i+j+1)}(K)]$$

$$\times \int_{nT}^{(n+1)T} [(n+1)T - t]p_i[H(t) - H(nT)]h(t)\,dt$$

$$= \sum_{n=0}^{\infty}\sum_{j=0}^{N-1} p_j[H(nT)]\sum_{i=0}^{\infty}[G^{(j)}(K) - G^{(i+j)}(K)]\int_{nT}^{(n+1)T} p_i[H(t) - H(nT)]\,dt,$$

$$(6.22)$$

where $(6.19) + (6.22) = (6.21)$. In this case, the expected cost rate is

$$\tilde{C}_2(N) = \frac{\begin{aligned} &c_N + (c_K - c_N)\sum_{n=0}^{\infty}\sum_{j=0}^{N-1} p_j[H(nT)] \\ &\times \sum_{i=0}^{\infty} p_i[H((n+1)T) - H(nT)][G^{(j)}(K) - G^{(i+j)}(K)] \\ &-c_D \sum_{n=0}^{\infty}\sum_{j=0}^{N-1} p_j[H(nT)]\sum_{i=0}^{\infty} G^{(i+j)}(K) \\ &\times \int_{nT}^{(n+1)T} p_i[H(t) - H(nT)]\,dt \end{aligned}}{T\sum_{n=0}^{\infty}\sum_{j=0}^{N-1} p_j[H(nT)]G^{(j)}(K)}$$

$$+ c_D \qquad (N = 1, 2, \cdots). \qquad (6.23)$$

It would be troublesome to analyze optimum policies analytically that minimize $C_2(N)$ and $\widetilde{C}_2(N)$. In particular, we derive an optimum shock number N^* that minimizes $\widetilde{C}_2(N)$ in (6.23) when $c_D = 0$ and $p_j[H(t)] = [(\lambda t)^j/j!]\mathrm{e}^{-\lambda t} = p_j(\lambda t)$ $(j = 0,1,2,\cdots)$. In this case, from the inequality $\widetilde{C}_2(N+1) - \widetilde{C}_2(N) \geq 0$,

$$\left[1 - \frac{\sum_{i=0}^{\infty} p_i(\lambda T)G^{(N+i)}(K)}{G^{(N)}(K)}\right] \sum_{n=0}^{\infty} \sum_{j=0}^{N-1} p_j(n\lambda T)G^{(j)}(K)$$

$$-\sum_{n=0}^{\infty} \sum_{j=0}^{N-1} p_j(n\lambda T) \sum_{i=0}^{\infty} p_i(\lambda T)[G^{(j)}(K) - G^{(i+j)}(K)] \geq \frac{c_N}{c_K - c_N}$$

$$(N = 1,2,\cdots). \tag{6.24}$$

Denoting the left-hand side of (6.24) by $Q(N)$,

$$Q(N+1) - Q(N) = \left[\frac{\sum_{i=0}^{\infty} p_i(\lambda T)G^{(N+i)}(K)}{G^{(N)}(K)} - \frac{\sum_{i=0}^{\infty} p_i(\lambda T)G^{(N+1+i)}(K)}{G^{(N+1)}(K)}\right]$$

$$\times \sum_{n=0}^{\infty} \sum_{j=0}^{N} p_j(n\lambda T)G^{(j)}(K).$$

Thus, if $\sum_{i=0}^{\infty} p_i(\lambda T)G^{(N+i)}(K)/G^{(N)}(K)$ is strictly decreasing in N and $Q(\infty) > c_N/(c_K - c_N)$, there exists a unique minimum number N^* $(1 \leq N^* < \infty)$ that satisfies (6.24).

For example, suppose that $G(x) = 1 - \mathrm{e}^{-\mu x}$, i.e., $G^{(j)}(x) \equiv \sum_{i=j}^{\infty} [(\mu x)^i/i!] \times \mathrm{e}^{-\mu x}$ $(j = 0,1,2,\cdots)$. Then,

$$\sum_{i=0}^{\infty} p_i(\lambda T)[G^{(N+i)}(K)G^{(N+1)}(K) - G^{(N+i+1)}(K)G^{(N)}(K)]$$

$$= \sum_{i=0}^{\infty} p_i(\lambda T)\mathrm{e}^{-2\mu K} \sum_{j=0}^{\infty} (\mu K)^{N+i+j} \left[\frac{1}{(N+i)!(N+1)!} - \frac{1}{N!(N+i+1)!}\right] > 0.$$

Thus, $\sum_{i=0}^{\infty} p_i(\lambda T)G^{(N+i)}(K)/G^{(N)}(K)$ is strictly decreasing to 0, and

$$Q(\infty) \equiv \lim_{N \to \infty} Q(N) = \sum_{n=1}^{\infty} \sum_{j=0}^{\infty} p_j(n\lambda T)G^{(j)}(K).$$

Therefore, if $\sum_{n=1}^{\infty} \sum_{j=0}^{\infty} p_j(n\lambda T)G^{(j)}(K) > c_N/(c_K - c_N)$, then an optimum N^* exists uniquely. Furthermore, from (6.11), if $(1 + \mu K)/(\lambda T) > c_K/(c_K - c_N)$, then a finite N^* exists.

(3) PM at Time NT

Suppose that we cannot know any damage level and shock number. The PM is done at time NT or the CM is done after failure, whichever occurs first, that is the same policy as that of Section 5.2. Then, the probability that the CM is done after failure is

$$\sum_{n=0}^{N-1}\sum_{j=0}^{\infty} p_j[H(nT)]\sum_{i=0}^{\infty} p_i[H((n+1)T) - H(nT)][G^{(j)}(K) - G^{(i+j)}(K)]$$

$$= \sum_{n=0}^{N-1}\sum_{j=0}^{\infty}\{p_j[H(nT)] - p_j[H((n+1)T)]\}G^{(j)}(K)$$

$$= 1 - \sum_{j=0}^{\infty} p_j[H(NT)]G^{(j)}(K), \tag{6.25}$$

and the probability that the PM is done at time NT is

$$\sum_{j=0}^{\infty} p_j[H(NT)]G^{(j)}(K). \tag{6.26}$$

The mean time to either PM or CM is

$$\sum_{n=0}^{N-1}\sum_{j=0}^{\infty} p_j[H(nT)]\sum_{i=0}^{\infty}[G^{(i+j)}(K) - G^{(i+j+1)}(K)]$$

$$\times \int_{nT}^{(n+1)T} t\, p_i[H(t) - H(nT)]h(t)\,\mathrm{d}t + (NT)\sum_{j=0}^{\infty} p_j[H(NT)]G^{(j)}(K)$$

$$= \sum_{n=0}^{N-1}\sum_{j=0}^{\infty} p_j[H(nT)]\sum_{i=0}^{\infty} G^{(i+j)}(K)\int_{nT}^{(n+1)T} p_i[H(t) - H(nT)]\,\mathrm{d}t$$

$$= \sum_{j=0}^{\infty} G^{(j)}(K)\int_0^{NT} p_j[H(t)]\,\mathrm{d}t. \tag{6.27}$$

Therefore, the expected cost rate is

$$C_3(N) = \frac{c_K - (c_K - c_N)\sum_{j=0}^{\infty} p_j[H(NT)]G^{(j)}(K)}{\sum_{j=0}^{\infty} G^{(j)}(K)\int_0^{NT} p_j[H(t)]\,\mathrm{d}t} \qquad (N = 1, 2, \cdots), \tag{6.28}$$

where c_N is the PM cost at time NT. The expected cost rate $C_3(N)$ agrees with $C_1(T)$ in (3.11) by replacing T with NT and $F^{(j)}(t) - F^{(j+1)}(t)$ with $p_j[H(t)]$.

Furthermore, when a failure level K is statistically distributed according to a general distribution $L(x)$, the expected cost rate is

$$C_3(N) = \frac{c_K - (c_K - c_N)\sum_{j=0}^{\infty} p_j[H(NT)]\int_0^{\infty} G^{(j)}(x)\,\mathrm{d}L(x)}{\sum_{j=0}^{\infty}\int_0^{NT} p_j[H(t)]\,\mathrm{d}t \int_0^{\infty} G^{(j)}(x)\,\mathrm{d}L(x)}$$

$$(N = 1, 2, \cdots). \tag{6.29}$$

Table 6.1. Optimum number N^* and expected cost rate $C_3(N^*)/c_N$ when $1/\lambda = 10^3$, 10^4, and $G^*(\theta) = 0.9$

T	$1/\lambda = 10^3$		$1/\lambda = 10^4$	
	N^*	$C_3(N^*)/c_N$	N^*	$C_3(N^*)/c_N$
8	13	0.0479	41	0.0151
48	2	0.0466	7	0.0153
192	1	0.0557	2	0.0159
2304	1	0.0564	1	0.0178

In particular, when $L(x) = 1 - e^{-\theta x}$, the expected cost rate is simplified as

$$C_3(N) = \frac{c_K - (c_K - c_N)e^{-[1-G^*(\theta)]H(NT)}}{\int_0^{NT} e^{-[1-G^*(\theta)]H(t)}\,dt} \qquad (N = 1, 2, \cdots), \qquad (6.30)$$

that agrees with (9.1) of [1] by replacing $\overline{F}(t)$ with $e^{-[1-G^*(\theta)]H(t)}$. Thus, when the failure rate $h(t)$ is strictly increasing, the optimum policy is as follows:

(i) If $h(\infty)[1 - G^*(\theta)] \int_0^\infty e^{-[1-G^*(\theta)]H(t)}\,dt > c_K/(c_K - c_N)$, then there exists a finite and unique minimum number N^* that satisfies

$$\frac{e^{-[1-G^*(\theta)]H(NT)} - e^{-[1-G^*(\theta)]H((N+1)T)}}{\int_{NT}^{(N+1)T} e^{-[1-G^*(\theta)]H(t)}\,dt} \int_0^{NT} e^{-[1-G^*(\theta)]H(t)}\,dt$$
$$- e^{-[1-G^*(\theta)]H(NT)} \geq \frac{c_K}{c_K - c_N}. \qquad (6.31)$$

(ii) If $h(\infty)[1 - G^*(\theta)] \int_0^\infty e^{-[1-G^*(\theta)]H(t)}\,dt \leq c_K/(c_K - c_N)$, then $N^* = \infty$, i.e., the unit is replaced only at failure and

$$C_3(\infty) = \frac{c_K}{\int_0^\infty e^{-[1-G^*(\theta)]H(t)}\,dt}.$$

Example 6.1. Suppose that $H(t) = \lambda t^2$, i.e., $h(t) = 2\lambda t$ that is strictly increasing to ∞. Thus, there exists a finite and unique minimum N^* that satisfies (6.31). Table 6.1 presents the optimum N^* and the resulting cost rate $C_3(N^*)/c_N$ for $T = 8, 48, 192, 2304$ when $c_K/c_N = 5$, $1/\lambda = 10^3$, 10^4, and $G^*(\theta) = 0.9$. For example, when $1/\lambda = 10^4$ and $T = 48$, i.e., the unit is operating 8 hours per day and is inspected once a week, the PM is done every 7 weeks. Clearly, optimum values of N^* decrease with T and increase with $1/\lambda$. ∎

If the failure is detected only at time nT $(n = 1, 2, \cdots)$, the mean time to either PM or CM is

$$\sum_{n=0}^{N-1}[(n+1)T]\sum_{j=0}^{\infty}p_j[H(nt)]\sum_{i=0}^{\infty}p_i[H((n+1)T) - H(nT)][G^{(j)}(K) - G^{(i+j)}(K)]$$

$$+ (NT)\sum_{j=0}^{\infty}p_j[H(NT)]G^{(j)}(K)$$

$$= T\sum_{n=0}^{N-1}\sum_{j=0}^{\infty}p_j[H(nT)]G^{(j)}(K), \qquad (6.32)$$

and the mean time from a failure to its detection is

$$\sum_{n=0}^{N-1}\sum_{j=0}^{\infty}p_j[H(nt)]\sum_{i=0}^{\infty}[G^{(i+j)}(K) - G^{(i+j+1)}(K)]$$

$$\times \int_{nT}^{(n+1)T}[(n+1)T - t]p_i[H(t) - H(nT)]h(t)\, dt$$

$$= \sum_{n=0}^{N-1}\sum_{j=0}^{\infty}p_j[H(nT)]\sum_{i=0}^{\infty}[G^{(j)}(K) - G^{(i+j)}(K)]\int_{nT}^{(n+1)T}p_i[H(t) - H(nT)]\, dt.$$

$$(6.33)$$

Thus, the expected cost rate is

$$\tilde{C}_3(N) = \frac{c_K - (c_K - c_N)\sum_{j=0}^{\infty}p_j[H(NT)]G^{(j)}(K) - c_D\sum_{j=0}^{\infty}G^{(j)}(K)\int_0^{NT}p_j[H(t)]\, dt}{T\sum_{n=0}^{N-1}\sum_{j=0}^{\infty}p_j[H(nT)]G^{(j)}(K)} + c_D \quad (N = 1, 2, \cdots),$$

$$(6.34)$$

where c_D is given in (6.12).

In addition, when a failure level K is distributed according to an exponential distribution $L(x) = 1 - e^{-\theta x}$, the expected cost rate is

$$\tilde{C}_3(N) = \frac{c_K - (c_K - c_N)e^{-[1-G^*(\theta)]H(NT)} - c_D\int_0^{NT}e^{-[1-G^*(\theta)]H(t)}\, dt}{T\sum_{n=0}^{N-1}e^{-[1-G^*(\theta)]H(nT)}} + c_D$$

$$(N = 1, 2, \ldots). \qquad (6.35)$$

7

Imperfect Preventive Maintenance Policies

The usual preventive maintenance (PM) of an operating unit is based on its age or operating time. Most models have assumed that the unit after PM becomes as good as new. Actually, this assumption might not be true. The unit after PM usually might be only younger, and its improvement would depend on the resources spent for PM. In such imperfect PM models where the unit after PM has the same failure rate as before PM, the age or failure rate after PM reduces in proportion to that before PM [212–214]. Some chapters [1,215–217] of recently published books summarized many results of imperfect maintenance.

The PM of large complex systems such as computers, radars, airplanes, and plants should be done frequently as the units age. A sequential PM policy where the PM is done at fixed intervals T_n ($n = 1, 2, \cdots, N$) has been proposed [218,219]. In some practical situations, however, the PM seems only imperfect in the sense that it does not make the unit like new [220].

In this chapter, we apply a sequential PM policy to a cumulative damage model where each PM is imperfect [57]: The unit is subject to shocks that occur randomly in time, and upon the occurrence of shocks, it suffers a random damage that is additive. Each shock causes unit failure with probability $p(x)$ when the total damage is x. If the unit fails between PMs, it undergoes only minimal repair using the same assumption as that of Section 5.4. We introduce only an improvement factor in damage to describe imperfect PM actions: The amount of damage after the nth PM becomes $a_n Z_n$ when it was Z_n before PM, $i.e.$, the nth PM reduces the total damage Z_n to $a_n Z_n$. This would be applied to related PM models in Chapter 6.

In Section 7.1, we obtain the expected cost rate when shocks occur in a Poisson process and $p(x)$ is exponential. In Section 7.2, we discuss three types of optimum policies that minimize the expected cost rate when the PM is done at periodic times and the improvement factor is constant, $i.e.$, $T_n = T$ and $a_n = a$. Optimum number $N^*(T)$, optimum interval $T^*(N)$, and optimum (N^*, T^*) are derived analytically. Numerical examples are presented to demonstrate potential usefulness of the results. Next, suppose in Section 7.4

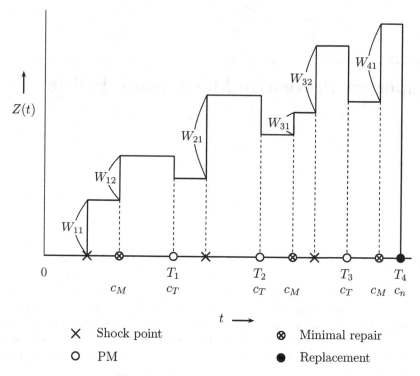

Fig. 7.1. Process for Imperfect PM

that a unit has to be operating over a finite interval $(0, S]$. Then, setting $\sum_{n=1}^{N} T_n = S$, we compute numerically an optimum number N^* and optimum times T_n^* $(n = 1, 2, \ldots, N^* - 1)$ that minimize the expected cost until replacement. It is of great interest that the last PM time interval is the largest and the first PM one is the second, and they are first increasing, and then are decreasing.

7.1 Model and Expected Cost

Consider a sequential PM policy that is done at fixed intervals T_n $(n = 1, 2, \cdots, N)$ and the replacement or the perfect PM is done at time T_N, *i.e.*, a unit is as good as new at time T_N. We call an interval from the $(n-1)$th PM to the nth PM *period* n (Figure 7.1).

Suppose that shocks occur in a Poisson process with rate λ. Random variables N_n $(n = 1, 2, \cdots, N)$ denote the number of shocks in period n, *i.e.*, $\Pr\{N_n = j\} = [(\lambda T_n)^j / j!] \exp(-\lambda T_n) \equiv p_j(T_n)$ $(j = 0, 1, 2, \cdots)$. In addition, we denote by W_{nj} the amount of damage caused by the jth shock in period n, where $W_{n0} \equiv 0$. It is assumed that random variable W_{nj} is nonnegative, independent, and identically distributed, and has an identical distribution $\Pr\{W_{nj} \le x\} \equiv G(x)$ for all n and j. The total damage is additive, and $G^{(j)}(x)$ $(j = 1, 2, \cdots)$ is the j-fold Stieltjes convolution of $G(x)$ with itself and $G^{(0)}(x) \equiv 1$ for all $x \ge 0$. Then, it follows that

$$\Pr\{W_{n1} + W_{n2} + \cdots + W_{nj} \le x\} = G^{(j)}(x) \qquad (j = 0, 1, 2, \cdots). \qquad (7.1)$$

When the total damage becomes x at shocks, the unit fails with probability $p(x)$, that is increasing in x from 0 to 1. If the unit fails between PMs, it undergoes only minimal repair, and hence, the total damage remains unchanged by any minimal repair. It is assumed that the times required for any PM and minimal repair are negligible.

Next, we introduce an improvement factor in PM: Suppose that the nth PM reduces $100(1 - a_n)\%$ $(0 \le a_n \le 1)$ of the total damage. Letting Z_n be the total damage at the end of period n, i.e., just before the nth PM, the nth PM reduces it to $a_n Z_n$. During period n the total damage is additive and is not removed because the failed unit undergoes only minimal repair. Thus, we have the relation

$$Z_n = a_{n-1} Z_{n-1} + \sum_{j=1}^{N_n} W_{nj} \qquad (n = 1, 2, \cdots, N), \qquad (7.2)$$

where $Z_0 \equiv 0$ and $\sum_{j=1}^{0} \equiv 0$.

Let c_T be the cost of each PM, c_N be the cost of replacement at the Nth PM with $c_N > c_T$, and c_M be the cost of minimal repair. Then, because the unit fails with probability $p(\cdot)$ only at shocks, the total cost in period n is

$$\widetilde{C}(n) = c_T + c_M \sum_{j=1}^{N_n} p(a_{n-1} Z_{n-1} + W_{n1} + W_{n2} + \cdots + W_{nj})$$

$$(n = 1, 2, \cdots, N - 1). \qquad (7.3)$$

Similarly, the total cost in period N is

$$\widetilde{C}(N) = c_N + c_M \sum_{j=1}^{N_N} p(a_{N-1} Z_{N-1} + W_{N1} + W_{N2} + \cdots + W_{Nj}). \qquad (7.4)$$

To obtain the expectations of (7.3) and (7.4), we assume that $p(x)$ is exponential, i.e., $p(x) = 1 - e^{-\theta x}$ for some constant $\theta > 0$. Letting $G^*(\theta)$ be the Laplace–Stieltjes transform of $G(x)$, i.e., $G^*(\theta) \equiv \int_0^\infty e^{-\theta x} dG(x)$,

$$E\{\exp[-\theta(W_{n1} + W_{n2} + \cdots + W_{nj})]\} = \int_0^\infty e^{-\theta x} dG^{(j)}(x) = [G^*(\theta)]^j.$$

$$(7.5)$$

The probability that the unit fails at the first shock is

$$\int_0^\infty p(x)\,dG(x) = \int_0^\infty (1 - e^{-\theta x})\,dG(x) = 1 - G^*(\theta).$$

Using the law of total probability in (7.3), the expected cost in period n is

$$E\left\{\widetilde{C}(n)\right\} = c_T + c_M E\left\{\sum_{j=1}^{N_n} p(a_{n-1}Z_{n-1} + W_{n1} + W_{n2} + \cdots + W_{nj})\right\}$$

$$= c_T + c_M \sum_{i=1}^\infty \Pr\left\{N_n = i\right\}$$

$$\times \sum_{j=1}^i E\left\{1 - \exp[-\theta(a_{n-1}Z_{n-1} + W_{n1} + W_{n2} + \cdots + W_{nj})]\right\}.$$

Let $B_n^*(\theta) \equiv E\left\{\exp(-\theta Z_n)\right\}$. Then, because Z_{n-1} and W_{nj} are independent of each other, from (7.5),

$$E\left\{1 - \exp[-\theta(a_{n-1}Z_{n-1} + W_{n1} + W_{n2} + \cdots + W_{nj})]\right\}$$
$$= 1 - B_{n-1}^*(\theta a_{n-1})[G^*(\theta)]^j.$$

Thus, from the assumption that N_n has a Poisson distribution with rate λ,

$$E\left\{\widetilde{C}(n)\right\} = c_T + c_M \sum_{k=1}^\infty \frac{(\lambda T_n)^k}{k!} e^{-\lambda T_n} \sum_{j=1}^k \left\{1 - B_{n-1}^*(\theta a_{n-1})[G^*(\theta)]^j\right\}$$

$$= c_T + c_M \left[\lambda T_n - \frac{G^*(\theta)}{1 - G^*(\theta)} B_{n-1}^*(\theta a_{n-1})\left\{1 - e^{-\lambda T_n[1 - G^*(\theta)]}\right\}\right]$$

$$(n = 1, 2, \cdots, N-1). \qquad (7.6)$$

Similarly, the expected cost in period N is

$$E\left\{\widetilde{C}(N)\right\} = c_N + c_M \left[\lambda T_N - \frac{G^*(\theta)}{1 - G^*(\theta)} B_{N-1}^*(\theta a_{N-1})\left\{1 - e^{-\lambda T_N[1 - G^*(\theta)]}\right\}\right].$$
$$(7.7)$$

It remains to determine $B_{n-1}^*(\theta a_{n-1})$. Let $A_j^n \equiv \prod_{i=j}^n a_i$ for $j \leq n$ and $\equiv 1$ for $j > n$. Then, from (7.2),

$$a_{n-1}Z_{n-1} = a_{n-1}a_{n-2}Z_{n-2} + a_{n-1}\sum_{i=1}^{N_{n-1}} W_{n-1i}$$

$$= \sum_{j=1}^{n-1}\left(A_j^{n-1}\sum_{i=1}^{N_j} W_{ji}\right),$$

so that,

$$B_{n-1}(\theta a_{n-1}) = E\left\{ e^{-\theta a_{n-1} Z_{n-1}} \right\} = E\left\{ \exp\left[-\theta \sum_{j=1}^{n-1} \left(A_j^{n-1} \sum_{i=1}^{N_j} W_{ji} \right) \right] \right\}.$$

Recalling that W_{ji} are independent and have an identical distribution $G(x)$,

$$E\left\{ \exp\left(-\theta A_j^{n-1} \sum_{i=1}^{N_j} W_{ji} \right) \right\} = \sum_{k=0}^{\infty} \Pr\{N_j = k\} E\left\{ \exp\left(-\theta A_j^{n-1} \sum_{i=1}^{k} W_{ji} \right) \right\}$$

$$= \sum_{k=0}^{\infty} \frac{(\lambda T_j)^k}{k!} e^{-\lambda T_j} [G^*(\theta A_j^{n-1})]^k$$

$$= \exp\left\{ -\lambda T_j [1 - G^*(\theta A_j^{n-1})] \right\},$$

and consequently,

$$B_{n-1}^*(\theta a_{n-1}) = \exp\left\{ -\sum_{j=1}^{n-1} \lambda T_j [1 - G^*(\theta A_j^{n-1})] \right\}. \tag{7.8}$$

Substituting (7.8) in (7.6) and (7.7), respectively, the expected costs in period n are

$$E\left\{ \widetilde{C}(n) \right\} = c_T + c_M \left[\lambda T_n - \frac{G^*(\theta)}{1 - G^*(\theta)} \exp\left\{ -\sum_{j=1}^{n-1} \lambda T_j [1 - G^*(\theta A_j^{n-1})] \right\} \right.$$

$$\left. \times \left\{ 1 - e^{-\lambda T_n [1 - G^*(\theta)]} \right\} \right] \qquad (n = 1, 2, \cdots, N-1), \tag{7.9}$$

and

$$E\left\{ \widetilde{C}(N) \right\} = c_N + c_M \left[\lambda T_N - \frac{G^*(\theta)}{1 - G^*(\theta)} \exp\left\{ -\sum_{j=1}^{N-1} \lambda T_j [1 - G^*(\theta A_j^{N-1})] \right\} \right.$$

$$\left. \times \left\{ 1 - e^{-\lambda T_N [1 - G^*(\theta)]} \right\} \right]. \tag{7.10}$$

Therefore, the expected cost rate until replacement is, from (7.9) and (7.10),

$$C_1(T_1, T_2, \cdots, T_N) = \frac{\sum_{n=1}^{N-1} E\{\widetilde{C}(n)\} + E\{\widetilde{C}(N)\}}{\sum_{n=1}^{N} T_n}$$

$$= \frac{(N-1)c_T + c_N + c_M \left[\sum_{n=1}^{N} \lambda T_n - G^*(\theta)/[1 - G^*(\theta)] \right.}{\sum_{n=1}^{N} T_n}$$

$$\frac{\times \sum_{n=1}^{N} \exp\left\{ -\sum_{j=1}^{n-1} \lambda T_j [1 - G^*(\theta A_j^{n-1})] \right\}}{\times \left\{ 1 - e^{-\lambda T_n [1 - G^*(\theta)]} \right\} \Big]}$$

$$(N = 1, 2, \cdots). \tag{7.11}$$

In the particular case of $N = 1$, $C_1(T_1)$ agrees with (5.27) by replacing c_T with c_N and $F(T) = 1 - e^{-\lambda T}$.

7.2 Optimum Policies

The expected cost rate $C_1(T_1, T_2, \cdots, T_N)$ in (7.11) is very complicated, and we cannot analyze optimum policies. Suppose that $T_n \equiv T$ and $a_n \equiv a$ ($0 \le a < 1$), i.e., the PM is done at periodic times nT ($n = 1, 2, \cdots, N$) and the improvement factor a_n is constant. Then, the expected cost rate is simplified as

$$C_1(N,T) = \lambda c_M + \frac{(N-1)c_T + c_N - c_M \{G^*(\theta)/[1 - G^*(\theta)]\} B_N(T)}{NT},$$

(7.12)

where

$$B_N(T) \equiv \left\{1 - e^{-\lambda[1-G^*(\theta)]T}\right\} \sum_{n=1}^{N} e^{-\lambda \xi_n T} \qquad (N = 1, 2, \cdots),$$

$$\xi_1 \equiv 0, \qquad \xi_n \equiv \sum_{j=1}^{n-1} [1 - G^*(\theta a^j)] \qquad (n = 2, 3, \cdots).$$

When $a = 0$, i.e., the PM is perfect, $\xi_n = 0$ and the expected cost rate is

$$C_1(N,T) = \lambda c_M$$
$$+ \frac{(N-1)c_T + c_N - N c_M \{G^*(\theta)/[1 - G^*(\theta)]\} \left\{1 - e^{-\lambda[1-G^*(\theta)]T}\right\}}{NT}. \quad (7.13)$$

The expected cost rate $C_1(N,T)$ in (7.13) is decreasing in N because $c_N > c_T$, and hence, $N^* = \infty$. Thus, an optimum interval T^* is easily derived by differentiating $C_1(\infty, T)$ and setting it equal to zero.

Before deriving optimum policies, we define a function that plays an important role in discussing them. Let

$$Q_n(T) \equiv c(n) - c_M \frac{G^*(\theta)}{1 - G^*(\theta)} \left\{1 - e^{-\lambda[1-G^*(\theta)]T}\right\} e^{-\lambda \xi_n T} \qquad (n = 1, 2, \cdots),$$

(7.14)

where $c(1) = c_N$ and $c(n) = c_T$ ($n = 2, 3, \cdots, N$). Then, (7.12) is rewritten as

$$C_1(N,T) = \lambda c_M + \frac{1}{NT} \sum_{n=1}^{N} Q_n(T).$$

(7.15)

(1) Optimum Number $N^*(T)$

We seek an optimum number $N^*(T)$ that minimizes $C_1(N,T)$ in (7.15) for a fixed $T > 0$ and $0 < a < 1$. From the inequality $C_1(N+1,T) \geq C_1(N,T)$,

$$L(N|T) \geq \frac{c_N - c_T}{c_N - Q_1(T)} \qquad (N = 1, 2, \cdots), \qquad (7.16)$$

where

$$L(N|T) \equiv \sum_{n=1}^{N} (e^{-\lambda \xi_n T} - e^{-\lambda \xi_{N+1} T}) \qquad (N = 1, 2, \cdots),$$

$$Q_1(T) = c_N - c_M \frac{G^*(\theta)}{1 - G^*(\theta)} \left\{ 1 - e^{-\lambda[1 - G^*(\theta)]T} \right\} < c_N.$$

Clearly,

$$L(N|T) - L(N-1|T) = N(e^{-\lambda \xi_N T} - e^{-\lambda \xi_{N+1} T}) > 0,$$

because ξ_n is strictly increasing in n. Thus, $L(N|T)$ is also strictly increasing in N.

Therefore, if $L(\infty|T) \equiv \lim_{N \to \infty} L(N|T) > (c_N - c_T)/[c_N - Q_1(T)]$, then there exists a finite and unique minimum $N^*(T)$ that satisfies (7.16).

Example 7.1. Suppose that the amount of damage at each shock has an exponential distribution $G(x) = 1 - e^{-\mu x}$ and $G^*(\theta) = \mu/(\theta + \mu)$. Then, $\xi_1 = 0$,

$$\xi_n = \sum_{j=1}^{n-1} \frac{a^j \theta}{a^j \theta + \mu} \qquad (n = 2, 3, \cdots).$$

It is assumed that the total damage is reduced in proportion to the PM cost c_T, *i.e.*, $c_T/c_N = 1 - a$. Table 7.1 presents the optimum number $N^*(T)$ and the resulting cost rate $C_1(N^*, T)/(\lambda c_M)$ for $a = 0.1$–0.9 and $c_N/c_M = 3$, 5, 10 when $\lambda T = 7$ and $G^*(\theta) = 0.9$, *i.e.*, $\mu/\theta = 9$. This indicates that $N^*(T)$ is not monotonically increasing with respect to a contrary to our expectation. However, this can be explained because $L(N|T)$ depends on a through c_T/c_N. For example, suppose that $T = 7$ days, *i.e.*, the PM is planned only on the weekend and shocks occur, on average, once a day. In this case, if $a = 0.5$ and $c_N/c_M = 5$, *i.e.*, both the costs of PM and minimal repair are half the replacement cost and the total damage is reduced to the half by PM, the unit should be replaced at three weeks. When a is small, several $N^*(T)$ become infinite. These cases show that the total damage is removed greatly by PM and the unit should undergo only PM rather than replacement. ∎

Table 7.1. Optimum number $N^*(T)$ and expected cost rate $C_1(N^*,T)/(\lambda c_M)$ when $G^*(\theta) = 0.9$, $\lambda T = 7$, and $c_T/c_N = 1 - a$

a	$c_N/c_M = 3$		$c_N/c_M = 5$		$c_N/c_M = 10$	
	$N^*(T)$	$C_1(N^*,T)/(\lambda c_M)$	$N^*(T)$	$C_1(N^*,T)/(\lambda c_M)$	$N^*(T)$	$C_1(N^*,T)/(\lambda c_M)$
0.9	2	0.7408	3	0.8917	7	1.1203
0.8	2	0.7508	3	0.9192	6	1.2084
0.7	2	0.7597	3	0.9443	6	1.2869
0.6	2	0.7674	3	0.9671	9	1.3569
0.5	2	0.7739	3	0.9876	∞^*	1.4086
0.4	2	0.7790	3	1.0062	∞	1.4656
0.3	1	0.7813	3	1.0229	∞	1.5324
0.2	1	0.7813	∞^*	1.0367	∞	1.6081
0.1	1	0.7813	∞	1.0487	∞	1.6915

∞^* indicates that $N^*(T)$ may not be infinite, but is very large.

(2) Optimum Number $T^*(N)$

We seek an optimum interval $T^*(N)$ that minimizes $C_1(N,T)$ in (7.15) for a fixed N. Differentiating $C_1(N,T)$ with respect to T and setting it equal to zero,

$$T \sum_{n=1}^{N} Q'_n(T) = \sum_{n=1}^{N} Q_n(T),$$

i.e.,

$$\sum_{n=1}^{N} \left[1 + \lambda T \xi_n - \{1 + \lambda T[1 - G^*(\theta) + \xi_n]\} - e^{-\lambda[1 - G^*(\theta)]T} \right] e^{-\lambda \xi_n T}$$

$$= \frac{(N-1)c_T + c_N}{c_M} \frac{1 - G^*(\theta)}{G^*(\theta)}. \tag{7.17}$$

When $n = 1$, $\xi_1 = 0$ and the term with $n = 1$ in the left-hand side of (7.17) is a gamma distribution of order 2, so that it increases from 0 to 1. The other terms with n ($n = 2, 3, \cdots, N$) are unimodal that is a unique solution of

$$e^{-\lambda[1 - G^*(\theta)]T} = \left(\frac{\xi_n}{1 - G^*(\theta) + \xi_n} \right)^2. \tag{7.18}$$

Thus, the left-hand side of (7.17) increases from 0 first, and then, oscillates and finally decreases to coverage to 1, as T increases. Therefore, there may be at most $(2N - 1)$ solutions that satisfy (7.17). An important $T^*(N)$ is either one of these solutions or $T^*(N) = \infty$. If there is no solution, then $T^*(N) = \infty$. In particular, when $N = 1$, there exists a unique solution that satisfies (7.17) if $G^*(\theta)/[1 - G^*(\theta)] > c_N/c_M$.

Table 7.2. Optimum time $T^*(N)$ and expected cost rate $C_1(N, T^*)/(\lambda c_M)$ when $G^*(\theta) = 0.9$, c_N/c_M, and $a = c_T/c_N = 0.5$

N	$T^*(N)$	$C_1(N, T^*)/(\lambda c_M)$
1	18.627	0.8603
2	13.358	0.9095
3	11.665	0.9429
4	10.816	0.9654
5	10.293	0.9811
6	9.933	0.9924
7	∞	1.0000

Example 7.2.   We compute $T^*(N)$ for $N = 1, 2, \cdots, 7$ when $G^*(\theta) = 0.9$, $c_N/c_M = 5$, and $a = c_T/c_N = 0.5$. Table 7.2 presents the values of $T^*(N)$ and $C_1(N, T)/(\lambda c_M)$ when N varies. In this case, the optimum interval becomes infinity for $N \geq 7$. ∎

(3) Optimum Pair (N^*, T^*)

We seek both optimum T^* and N^* that minimize $C_1(N, T)$ in (7.15). From (7.12), we can see that $C_1(N, \infty) = \lambda c_M$ for all $N \geq 1$. Thus, optimum (N^*, T^*) must satisfy $C_1(N^*, T^*) \leq \lambda c_M$. It follows from (7.12) that a necessary condition for (N^*, T^*) is that $Q_n(T^*) < 0$ for at least one $n \leq N^*$ because otherwise no contribution to the second term in (7.12) occurs.

Now, consider the inequality $Q_n(T) \leq 0$. This is equivalent to considering

$$h_n(T) \geq \frac{c(n)}{c_M} \frac{G^*(\theta)}{1 - G^*(\theta)}, \tag{7.19}$$

where

$$h_n(T) \equiv \left\{ 1 - e^{-\lambda[1 - G^*(\theta)]T} \right\} e^{-\lambda \xi_n T} \qquad (n = 1, 2, \cdots, N).$$

It is easy to see that $dh_n(T)/dT = 0$ has a unique solution m_n that satisfies

$$[1 - G^*(\theta) + \xi_n] e^{-\lambda[1 - G^*(\theta)]T} = \xi_n. \tag{7.20}$$

Thus, $h_n(T)$ is unimodal with m_m, and hence,

$$h_n(T) \leq h_n(m_n)$$

$$= \left[1 - \frac{\xi_n}{1 - G^*(\theta) + \xi_n} \right] \left[\frac{\xi_n}{1 - G^*(\theta) + \xi_n} \right]^{\xi_n/[1 - G^*(\theta)]} < 1.$$

It is proved that both m_n and $h_n(m_n)$ are decreasing in n, so that both m_∞ and $h_\infty(m_\infty)$ exist. Thus, it follows that

$$N^* < n^* = \min_{n \geq 2} \left\{ h_n(m_n) \leq \frac{c_T}{c_M} \frac{1 - G^*(\theta)}{G^*(\theta)} \right\}. \qquad (7.21)$$

Here, if $h_\infty(m_\infty) > (c_T/c_M)[1 - G^*(\theta)]/G^*(\theta)$, then we set $N^* = \infty$. It can be seen that $T^* \geq m_{n^*-1}$ because m_n is decreasing in n. On the other hand, $T^* \leq \max\{T^*(1), m_2\}$. To this end, suppose that T satisfies (7.17), and recall that $Q'_n(T) < 0$ for $T < m_n$, $Q'_n \geq 0$ for $T \geq m_n$, and m_n is decreasing in n. Then, if $T^*(1) > m_2$, either $T^* = T^*(1)$ with $N^* = 1$ or $T^* < T^*(1)$. If $T^*(1) < m_2$, $T^* > m_2$ never happens because $\sum_{n=1}^{N} Q'_n(T^*)/N > Q'_1(T^*(1))$. Thus, $T^* \leq \max\{T^*(1), m_2\}$, as desired.

From the above analysis, we have the following optimum policy: Suppose that $n^* < \infty$ that is given in (7.20). Then, the optimum pair (N^*, T^*) is confined, as $N^* < n^*$ and $m_{n^*-1} \leq T^* \leq \max\{T^*(1), m_2\}$, where m_n is a unique solution of (7.20). Therefore, the optimum pair is given by

$$T^*(N^*) = \min_{1 \leq N \leq n^*} T^*(N) = \min_{m_{n^*-1} \leq T \leq \max\{T^*(1), m_2\}} N^*(T). \qquad (7.22)$$

Example 7.3. Consider the model in Example 7.2 and compute an optimum pair (N^*, T^*) that minimizes $C_1(N, T)$. In this example, $h_4(m_4) \approx 0.2621 < 0.27$, and hence, $N^* \leq 3$. In fact, Table 7.2 indicates that $N^* = 1$ and $T^* = 18.627$. ∎

7.3 Optimum Policies for a Finite Interval

Suppose that a unit has to be operating over a finite interval $(0, S]$ and be replaced at time S (Section 9.2 of [1]). When $a_n \equiv a$ and $G(x) = 1 - e^{-\mu x}$, $C_1(T_1, T_2, \ldots, T_N)$ is, from (7.11),

$$
\begin{aligned}
C_2(T_1, T_2, \ldots, T_{N-1}) &= c_M - \frac{C_1(T_1, T_2, \ldots, T_N)}{\lambda} \\
&= \frac{c_M(\mu/\theta)\sum_{n=1}^{N}\exp\left[-\sum_{j=1}^{n-1}\lambda A_{n-j}(\theta)T_j\right]}{\lambda \sum_{n=1}^{N} T_n} \\
&\quad \times [1 - e^{-\lambda A_0(\theta)T_n}] - (N-1)c_T - c_N
\end{aligned}
$$

$$(N = 1, 2, \ldots), \qquad (7.23)$$

where $T_1 + T_2 + \cdots + T_N = S$ and

$$A_j(\theta) \equiv \frac{\theta a^j}{\theta a^j + \mu} \qquad (j = 0, 1, 2, \ldots).$$

It is noted that $A_j(\theta) > A_{j+1}(\theta)$ $(j = 0, 1, 2, \ldots)$ for $0 < a < 1$. In this case, we consider the optimum policy that maximizes the expected cost

$$\widetilde{C}_2(T_1, T_2, \ldots, T_{N-1}) = \frac{\mu c_M}{\theta} \sum_{n=1}^{N} \exp\left[-\sum_{j=1}^{n-1} \lambda A_{n-j}(\theta) T_j\right] \left[1 - e^{-\lambda A_0(\theta) T_n}\right]$$
$$- (N-1)c_T - c_N \quad (N = 1, 2, \ldots). \quad (7.24)$$

For example, when $N = 1$, $i.e.$, no PM is done,

$$\widetilde{C}_2 = \frac{\mu c_M}{\theta} \left[1 - e^{-\lambda A_0(\theta) S}\right] - c_N, \quad (7.25)$$

that is constant.

When $N = 2$,

$$\widetilde{C}_2(T_1) = \frac{\mu c_M}{\theta} \left\{1 - e^{-\lambda A_0(\theta) T_1} + e^{-\lambda A_1(\theta) T_1} \left[1 - e^{-\lambda A_0(\theta)(S - T_1)}\right]\right\}$$
$$- c_T - c_N. \quad (7.26)$$

Differentiating $\widetilde{C}_2(T_1)$ with respect to T_1 and setting it equal to zero,

$$A_0(\theta) \left\{e^{-\lambda[A_0(\theta) - A_1(\theta)]T_1} - e^{-\lambda A_0(\theta)(S - T_1)}\right\} - A_1(\theta)\left[1 - e^{-\lambda A_0(\theta)(S - T_1)}\right] = 0. \quad (7.27)$$

Letting $Q(T_1)$ be the left-hand side of (7.27),

$$Q(0) = [A_0(\theta) - A_1(\theta)] \left[1 - e^{-\lambda A_0(\theta) S}\right] > 0,$$
$$Q(S) = -A_0(\theta) \left\{1 - e^{-\lambda[A_0(\theta) - A_1(\theta)]S}\right\} < 0,$$
$$Q'(T_1) = -A_0(\theta) [A_0(\theta) - A_1(\theta)] \left\{e^{-\lambda[A_0(\theta) - A_1(\theta)]T_1} + e^{-\lambda A_0(\theta)(S - T_1)}\right\} < 0.$$

Thus, there exists an optimum time T_1^* $(0 < T_1^* < S)$ that satisfies (7.27).

When $N = 3$,

$$\widetilde{C}_2(T_1, T_2) = \frac{\mu c_M}{\theta} \left\{1 - e^{-\lambda A_0(\theta) T_1} + e^{-\lambda A_1(\theta) T_1} \left[1 - e^{-\lambda A_0(\theta) T_2}\right]\right.$$
$$\left. + e^{-\lambda A_2(\theta) T_1 - \lambda A_1(\theta) T_2} \left[1 - e^{-\lambda A_0(\theta)(S - T_1 - T_2)}\right]\right\}$$
$$- 2c_T - c_N. \quad (7.28)$$

Differentiating $\widetilde{C}_2(T_1, T_2)$ with respect to T_1 and T_2 and setting them equal to zero, respectively,

$$A_0(\theta) \left[e^{-\lambda A_0(\theta) T_1} - e^{-\lambda A_2(\theta) T_1 - \lambda A_1(\theta) T_2 - \lambda A_0(\theta)(S - T_1 - T_2)}\right]$$
$$- A_1(\theta) e^{-\lambda A_1(\theta) T_1} \left[1 - e^{-\lambda A_0(\theta) T_2}\right]$$
$$- A_2(\theta) e^{-\lambda A_2(\theta) T_1 - \lambda A_1(\theta) T_2} \left[1 - e^{-\lambda A_0(\theta)(S - T_1 - T_2)}\right] = 0, \quad (7.29)$$

Table 7.3. PM times λT_n and expected cost $\tilde{C}_2(T_1, T_2, \ldots, T_{N-1})/c_M$ for $N = 1, 2, \ldots, 10$ when $a = 0.5$, $\mu/\theta = 10$, $c_N/c_M = 5$, $c_T/c_M = 1.0$, and $\lambda S = 40$

	$N=1$	$N=2$	$N=3$	$N=4$	$N=5$	$N=6$	$N=7$	$N=8$	$N=9$	$N=10$
λT_1	40.00	13.17	12.41	11.37	10.32	9.36	8.52	7.80	7.17	6.63
λT_2		26.83	5.60	5.27	4.82	4.38	3.99	3.66	3.37	3.11
λT_3			21.99	5.23	4.87	4.45	4.06	3.72	3.42	3.17
λT_4				18.14	4.78	4.45	4.07	3.73	3.44	3.18
λT_5					15.22	4.35	4.06	3.73	3.44	3.18
λT_6						13.01	3.97	3.71	3.44	3.18
λT_7							11.33	3.64	3.42	3.18
λT_8								10.01	3.35	3.16
λT_9									8.96	3.10
λT_{10}										8.10
$\frac{\tilde{C}_2(\cdot)}{c_M}$	4.74	5.86	6.87	7.70	8.34	8.78	9.05	9.17	9.16	9.03

$$A_0(\theta)\left[e^{-\lambda A_1(\theta)T_1 - \lambda A_0(\theta)T_2} - e^{-\lambda A_2(\theta)T_1 - \lambda A_1(\theta)T_2 - \lambda A_0(\theta)(S-T_1-T_2)} \right]$$
$$- A_1(\theta)e^{-\lambda A_2(\theta)T_1 - \lambda A_1(\theta)T_2}\left[1 - e^{-\lambda A_0(\theta)(S-T_1-T_2)} \right] = 0. \qquad (7.30)$$

In general, differentiating $\tilde{C}_2(T_1, T_2, \ldots, T_{N-1})$ with respect to T_n $(n = 1, 2, \ldots, N-1)$ $(N \geq 2)$ and setting them equal to zero,

$$A_0(\theta)\left\{ \exp\left[-\sum_{j=1}^{n} \lambda A_{n-j}(\theta)T_j \right] - \exp\left[-\sum_{j=1}^{N} \lambda A_{N-j}(\theta)T_j \right] \right\}$$
$$- \sum_{i=n+1}^{N} A_{i-n}(\theta)\left\{ \exp\left[-\sum_{j=1}^{i-1} \lambda A_{i-j}(\theta)T_j \right] - \exp\left[-\sum_{j=1}^{i} \lambda A_{i-j}(\theta)T_j \right] \right\} = 0$$
$$(n = 1, 2, \ldots, N-1), \qquad (7.31)$$

where note that $T_N = S - T_1 - T_2 - \cdots - T_{N-1}$.

Therefore, we may solve the simultaneous equations (7.31) and obtain the expected cost $\tilde{C}_2(T_1, T_2, \ldots, T_{N-1})$ in (7.24). Next, compared $\tilde{C}_2(T_1, T_2, \ldots, T_{N-1})$ for all $N \geq 1$, we can get the optimum number N^* and times T_n^* $(n = 1, 2, \ldots, N^* - 1)$ for a specified S.

Example 7.4. Table 7.3 presents λT_n $(n = 1, 2, \ldots, N)$ and $\tilde{C}_2(T_1, T_2, \ldots, T_{N-1})/c_M$ when $a = 0.5$, $\mu/\theta = 10$, $c_N/c_M = 5$, $c_T/c_M = 1.0$, and $\lambda \sum_{n=1}^{N} T_n = 40$ for $N = 1, 2, \ldots, 10$. Compared $\tilde{C}_2(T_1, T_2, \ldots, T_{N-1})$ for $N = 1, 2, \ldots, 10$, the expected cost $\tilde{C}_2(\cdot)$ is maximum, i.e., $C_2(\cdot)$ in (7.23) is minimum at $N^* = 8$. In this case, the optimum PM number is $N^* = 8$ and optimum PM times are 7.80, 11.46, 15.18, 18.91, 22.64, 26.35, 29.99, 40. This indicates the interesting result that the last PM time interval is the largest and the first one is the second, and they are first increasing, remain in constant for some number, and then decreasing for large N, that is, PM time intervals draw a upside-down bathtub curve [221] for $2 \leq n \leq N-1$. PM interval times T_n ($n = 1, 2, \ldots, 10$), draws roughly a standard bathtub curve. It would be necessary to inquire into why the PM time intervals describe the two bathtub curves. ∎

8

Garbage Collection Policies

A database for a computer system is in optimum storage according to the scheme defined in the data structures. However, after some operations, storage areas are not in good order due to additions and deletions of data. Such updating procedures reduce the size of continuous and available memory areas, and make processing efficiency worse. To use storage areas effectively and to improve processing efficiently, garbage collections (GCs) have to be done at suitable times. Many GCs to reclaim the storage and rearrange a database are used in most large list processing systems [222, 223]. Some algorithms for performing the GC of linked data structures were reviewed [224]. Several authors have studied real time GCs to avoid suspension of the application program in its execution [225–227]. Most problems have been concerned with ways to introduce GC methods.

When a database is updated from several online terminals, it is necessary to set up a desired response time. If response times become comparatively long, the processing efficiency becomes worse, and finally, it would be impossible to update data. Such response times may depend on the amount of garbage in a database.

This chapter proposes when to make the GC for a database with an upper limit level K of the total garbage. An amount of garbage with a general distribution $G(x)$ arises from each update and is additive. A cost and time for the GC are higher if the total garbage is greater than K. In Section 8.1, to prevent such the event, the GC is done at periodic time T or at the Nth update, whichever occurs first [58]. It is assumed in Section 8.2 that if there exist data that are not erased, they remain in the storage area as garbage. In Section 8.3, a database is checked at periodic times to investigate the amount of garbage. If the total garbage exceeds a managerial level Z, the GC is done. Using the results of Section 6.1, the optimum policy is derived. Each GC restores computer resources such as response time, storage area, and throughput to an initial state. This corresponds to one modification of maintenance policies for cumulative damage models, replacing *update* with *shock* and *garbage* with *damage*. Using the results of Chapters 3, 5, and 6, the expected cost rates or the availabilities are derived, and optimum policies that minimize them are discussed analytically. Numerical examples are given when a database is updated in a Poisson process and an amount of garbage due to updates is exponential. It is theoretically noted that the policy maximizing the availability corresponds essentially to the policy minimizing the expected cost rate.

8.1 Standard Garbage Collection Model

Suppose that a database is updated in a nonhomogeneous Poisson process with an intensity function $h(t)$ and a mean value function $H(t)$, i.e., $H(t) \equiv \int_0^t h(u)\mathrm{d}u$. Then, the probability of j updates in $[0,t]$ is $p_j(t) \equiv \{[H(t)]^j/j!\}\,\mathrm{e}^{-H(t)}$ $(j = 0, 1, 2, \cdots)$. Furthermore, an amount W_j of garbage arises from the jth update and has a probability distribution $G(x) \equiv \Pr\{W_j \le x\}$, independent of the number of updates, and these amount of garbage are additive. Then, the total garbage $\sum_{i=1}^j W_i$ up to the jth update has $\Pr\{\sum_{i=1}^j W_i \le x\} = G^{(j)}(x)$ $(j = 1, 2, \cdots)$, where $G^{(j)}(x)$ is the j-fold Stieltjes convolution of $G(x)$ with itself and $G^{(0)}(x) \equiv 1$ for $x \ge 0$. When the total garbage has exceeded an upper limit level K, the database becomes useless for lack of storage area or due to a long response time.

To prevent the database becoming useless, the GC is done at a planned time T or at an update number N, whichever occurs first. For the above model, we introduce the following costs: c_T and c_N are the fixed costs for the respective GCs at time T and update N, and c_K is the fixed cost for the GC when the total garbage has exceeded a level K with $c_K > c_T$ and $c_K > c_N$. In addition, $c_0(x)$ is a variable cost for the collection of an amount x $(0 \le x \le K)$ of garbage.

Using a method similar to (1) of Section 3.3, the expected cost when the GC is done at time T or at update N is

$$\sum_{j=0}^{N-1} p_j(T) \int_0^K [c_T + c_0(x)]\,\mathrm{d}G^{(j)}(x)$$

$$+ \int_0^T p_{N-1}(t)h(t)\,\mathrm{d}t \int_0^K [c_N + c_0(x)]\,\mathrm{d}G^{(N)}(x), \tag{8.1}$$

and the expected cost when the total garbage has exceeded a level K is

$$[c_K + c_0(K)] \sum_{j=0}^{N-1} [G^{(j)}(K) - G^{(j+1)}(K)] \int_0^T p_j(t)h(t)\,\mathrm{d}t. \tag{8.2}$$

The mean time to GC is

$$T \sum_{j=0}^{N-1} p_j(T) G^{(j)}(K) + G^{(N)}(K) \int_0^T t \, p_{N-1}(t) h(t) \, dt$$

$$+ \sum_{j=0}^{N-1} [G^{(j)}(K) - G^{(j+1)}(K)] \int_0^T t \, p_j(t) h(t) \, dt = \sum_{j=0}^{N-1} G^{(j)}(K) \int_0^T p_j(t) \, dt.$$

$$(8.3)$$

Therefore, the expected cost rate is, summing up (8.1) and (8.2), and dividing by (8.3),

$$C(T,N) = \frac{\begin{aligned}&\sum_{j=0}^{N-1} p_j(T) \int_0^K [c_T + c_0(x)] \, dG^{(j)}(x) \\ &+ \int_0^T p_{N-1}(t) h(t) \, dt \int_0^K [c_N + c_0(x)] \, dG^{(N)}(x) \\ &+ [c_K + c_0(K)] \sum_{j=0}^{N-1} [G^{(j)}(K) - G^{(j+1)}(K)] \int_0^T p_j(t) h(t) \, dt\end{aligned}}{\sum_{j=0}^{N-1} G^{(j)}(K) \int_0^T p_j(t) \, dt},$$

$$(8.4)$$

and

$$C(\infty) \equiv \lim_{\substack{T \to \infty \\ N \to \infty}} C(T,N)$$

$$= \frac{c_K + c_0(K)}{\sum_{j=0}^{\infty} G^{(j)}(K) \int_0^{\infty} p_j(t) \, dt},$$

$$(8.5)$$

(1) Optimum T^*

Suppose that the GC is done only at time T. Then, from (8.4), the expected cost rate is given by

$$C_1(T) \equiv \lim_{N \to \infty} C(T,N)$$

$$= \frac{\begin{aligned}&\sum_{j=0}^{\infty} p_j(T) \int_0^K [c_T + c_0(x)] \, dG^{(j)}(x) \\ &+ [c_K + c_0(K)] \sum_{j=0}^{\infty} [G^{(j)}(K) - G^{(j+1)}(K)] \int_0^T p_j(t) h(t) \, dt\end{aligned}}{\sum_{j=0}^{\infty} G^{(j)}(K) \int_0^T p_j(t) \, dt}.$$

$$(8.6)$$

We seek an optimum time T^* that minimizes $C_1(T)$ in (8.6) when $c_0(x) = c_0 x$. Differentiating $C_1(T)$ with respect to T and setting it equal to zero,

$$(c_K - c_T) \left\{ h(T) Q_1(T) \sum_{j=0}^{\infty} G^{(j)}(K) \int_0^T p_j(t) \, dt - \sum_{j=0}^{\infty} p_j(T)[1 - G^{(j)}(K)] \right\}$$

$$+ c_0 \left\{ h(T) Q_2(T) \sum_{j=0}^{\infty} G^{(j)}(K) \int_0^T p_j(t) \, dt - \sum_{j=0}^{\infty} p_j(T) \int_0^K [1 - G^{(j)}(x)] \, dx \right\}$$

$$= c_T,$$

$$(8.7)$$

where

$$Q_1(T) = \frac{\sum_{j=0}^{\infty} p_j(T)[G^{(j)}(K) - G^{(j+1)}(K)]}{\sum_{j=0}^{\infty} p_j(T)G^{(j)}(K)},$$

$$Q_2(T) = \frac{\sum_{j=0}^{\infty} p_j(T) \int_0^K [G^{(j)}(x) - G^{(j+1)}(x)]\, dx}{\sum_{j=0}^{\infty} p_j(T)G^{(j)}(K)}.$$

In the particular case of $c_0 = 0$, (8.7) becomes

$$h(T)Q_1(T)\sum_{j=0}^{\infty} G^{(j)}(K) \int_0^T p_j(t)\, dt - \sum_{j=0}^{\infty} p_j(T)[1 - G^{(j)}(K)] = \frac{c_T}{c_K - c_T}.$$

$$(8.8)$$

If $h(T)Q_1(T)$ is strictly increasing, then the left-hand side of (8.8) is also strictly increasing in T from 0 to $h(\infty)Q_1(\infty)\sum_{j=0}^{\infty} G^{(j)}(K)\int_0^{\infty} p_j(t)dt - 1$, where $h(\infty) \equiv \lim_{t \to \infty} h(t)$ and $Q_1(\infty) \equiv \lim_{t \to \infty} Q_1(t)$. Thus, if

$$h(\infty)Q_1(\infty)\sum_{j=0}^{\infty} G^{(j)}(K) \int_0^{\infty} p_j(t)\, dt > \frac{c_K}{c_K - c_T},$$

then there exists a finite and unique T^* that satisfies (8.8).

In addition, when $p_j(t) = [(\lambda t)^j/j!]e^{-\lambda t}$ and $G^{(j)}(x) = \sum_{i=j}^{\infty}[(\mu x)^i/i!]e^{-\mu x}$ $(j = 0, 1, 2, \cdots)$, (8.7) is simplified as

$$Q_1(T)\sum_{j=0}^{\infty} G^{(j)}(K) \sum_{i=j+1}^{\infty} p_i(T) - \sum_{j=0}^{\infty}[1 - G^{(j)}(K)]p_j(T) = \frac{c_T}{c_K - c_T - c_0/\mu},$$

$$(8.9)$$

that agrees with (3.34). Thus, if $c_K > c_T[1 + (1/\mu K)] + c_0/\mu$, then there exists a finite and unique T^* that satisfies (8.9), and the resulting cost rate is given in (3.36). Conversely, if $c_K \le c_T[1 + (1/\mu K)] + c_0/\mu$, then $T^* = \infty$, and the resulting cost rate is given in (8.5).

(2) Optimum N^*

The expected cost rate when the GC is done only at update N is, from (8.4),

$$C_2(N) \equiv \lim_{T \to \infty} C(T, N)$$

$$= \frac{[c_K + c_0(K)][1 - G^{(N)}(K)] + \int_0^K [c_N + c_0(x)]\, dG^{(N)}(x)}{\sum_{j=0}^{N-1} G^{(j)}(K) \int_0^{\infty} p_j(t)\, dt}$$

$$(N = 1, 2, \dots). \qquad (8.10)$$

Forming the inequality $C_2(N + 1) - C_2(N) \ge 0$ to seek an optimum number N^* that minimizes $C_2(N)$ in (8.10) when $c_0(x) = c_0 x$,

$$(c_K - c_N)\left\{\frac{G^{(N)}(K) - G^{(N+1)}(K)}{G^{(N)}(K)\int_0^\infty p_N(t)\,dt}\sum_{j=0}^{N-1}G^{(j)}(K)\int_0^\infty p_j(t)\,dt - [1 - G^{(N)}(K)]\right\}$$

$$+ c_0\left\{\frac{\int_0^K [G^{(N)}(x) - G^{(N+1)}(x)]\,dx}{G^{(N)}(K)\int_0^\infty p_N(t)\,dt}\sum_{j=0}^{N-1}G^{(j)}(K)\int_0^\infty p_j(t)\,dt\right.$$

$$\left. - \int_0^K [1 - G^{(N)}(x)]\,dx\right\} \geq c_N \qquad (N = 1, 2, \dots). \tag{8.11}$$

When $c_0 = 0$ and $p_j(t) = [(\lambda t)^j/j!]e^{-\lambda t}$, (8.11) is

$$Q_3(N)\sum_{j=0}^{N-1}G^{(j)}(K) - [1 - G^{(N)}(K)] \geq \frac{c_N}{c_K - c_N} \qquad (N = 1, 2, \cdots), \tag{8.12}$$

that agrees with (3.22) where $Q_3(N) \equiv [G^{(N)}(K) - G^{(N+1)}(K)]/G^{(N)}(K)$ and represents the discrete failure rate defined in (2.15). Thus, if $Q_3(N)$ is strictly increasing and $Q_3(\infty)[1 + M_G(K)] > c_K/(c_K - c_N)$, where $M_G(K) \equiv \sum_{j=1}^\infty G^{(j)}(K)$, then there exists a finite and unique minimum N^* ($1 \leq N^* < \infty$) that satisfies (8.12). In addition, when $G^{(j)}(x) = \sum_{i=j}^\infty [(\mu x)^i/i!]e^{-\mu x}$, $Q_3(N)$ is strictly increasing from $e^{-\mu K}$ to 1 from Example 2.2 of Chapter 2. Thus, if $\mu K > c_N/(c_K - c_N)$, then there exists a finite and unique minimum N^* that satisfies (8.12).

Example 8.1. We compute optimum T^* and N^* when $c_0(x) = c_0 x$, $h(t) = \lambda$ and $G(x) = 1 - e^{-\mu x}$. Under such assumptions, (8.9) and (8.11) are rewritten as, respectively,

$$\frac{\sum_{j=0}^\infty [(\lambda T)^j/j!][(\mu K)^j/j!]}{\sum_{j=0}^\infty [(\lambda T)^j/j!]\sum_{i=j}^\infty [(\mu K)^i/i!]}$$

$$\times \sum_{j=0}^\infty\left\{\sum_{i=j+1}^\infty [(\lambda T)^i/i!]e^{-\lambda T}\right\}\left\{\sum_{i=j}^\infty [(\mu K)^i/i!]e^{-\mu K}\right\}$$

$$- \sum_{j=1}^\infty \frac{(\lambda T)^j}{j!}e^{-\lambda T}\sum_{i=0}^{j-1}\frac{(\mu K)^i}{i!}e^{-\mu K} = \frac{c_T}{c_K - c_T - c_0/\mu}, \tag{8.13}$$

and

$$\frac{[(\mu K)^N/N!]}{\sum_{j=N}^\infty [(\mu K)^j/j!]}\sum_{j=0}^{N-1}\sum_{i=j}^\infty \frac{(\mu K)^i}{i!}e^{-\mu K} - \sum_{j=0}^{N-1}\frac{(\mu K)^j}{j!}e^{-\mu K} \geq \frac{c_N}{c_K - c_N - c_0/\mu}. \tag{8.14}$$

If $c_K > c_k[1 + (1/\mu K)] + c_0/\mu$ ($k = T, N$), then there exist both finite T^* and N^* that satisfies (8.13) and (8.14), respectively.

Table 8.1. Optimum time λT^* and expected cost rate $C_1(T^*)/(\lambda c_T)$ when $c_0 K/c_T = 1$

c_K/c_T	$\mu K = 150$		$\mu K = 300$	
	λT^*	$C_1(T^*)/(\lambda c_T) \times 10^2$	λT^*	$C_1(T^*)/(\lambda c_T) \times 10^3$
100	98.1	1.715	221.5	7.904
200	95.3	1.734	217.5	8.026
500	92.0	1.790	212.5	8.115
1000	89.6	1.808	209.0	8.223

c_K/c_T	$\mu K = 500$		$\mu K = 700$	
	λT^*	$C_1(T^*)/(\lambda c_T) \times 10^3$	λT^*	$C_1(T^*)/(\lambda c_T) \times 10^3$
100	394.5	4.576	572.1	3.191
200	389.2	4.614	565.8	3.213
500	382.6	4.643	558.0	3.244
1000	377.9	4.663	552.4	3.259

Table 8.2. Optimum number N^* and expected cost rate $C_2(N^*)/(\lambda c_N)$ when $c_0 K/c_N = 1$

c_K/c_N	$\mu K = 150$		$\mu K = 300$	
	N^*	$C_2(N^*)/(\lambda c_N) \times 10^2$	N^*	$C_2(N^*)/(\lambda c_N) \times 10^3$
100	110	1.600	241	7.562
200	108	1.617	238	7.613
500	105	1.640	234	7.678
1000	103	1.657	231	7.725

c_K/c_N	$\mu K = 500$		$\mu K = 700$	
	N^*	$C_2(N^*)/(\lambda c_N) \times 10^3$	N^*	$C_2(N^*)/(\lambda c_N) \times 10^3$
100	421	4.406	605	3.100
200	417	4.428	600	3.112
500	412	4.455	594	3.127
1000	409	4.475	590	3.139

Table 8.1 presents the optimum T^* for $\mu K = 150, 300, 500, 700$ and $c_K/c_T = 100, 200, 500, 1000$ when $c_0 K/c_T = 1$, *i.e.*, $c_0/\mu = c_T/(\mu K)$. In this case, if $c_K/c_T > 1 + (2/\mu K)$, then a finite T^* exists. For example, when $\lambda = 5$, $c_K/c_T = 100$, and $\mu K = 700$, the optimum time is $\lambda T^* = 572.1$. This indicates that when the database is updated 5 times an hour and becomes useless after 700 updates, on average, the GC should be done at $572.1/5 = 114.42$ hour, *i.e.*, at about $114.42/24 \approx 4.8$ days. Taking another viewpoint, when the total garbage has exceeded $(572.1/700) \times 100 \approx 81.7\%$ of an upper limit μK, the GC should be done.

Similarly, Table 8.2 presents the optimum number N^* for $\mu K = 150, 300,$ 500, 700 and $c_K/c_N = 100, 200, 500, 1000$ when $c_0 K/c_N = 1$. For example, when $c_K/c_N = 100$ and $\mu K = 700$, the optimum number is $N^* = 605$, that is, the GC is done at $(600/700) \times 100 \approx 86.4\%$ of an upper limit μK, whose values are greater than those, and the resulting cost rates are smaller than those in Table 8.1 when $c_T = c_N$. In this case, the GC policy at update N is more economical than that at time T, however, they have almost the same values. Furthermore, it is of interest that both T^* and N^* depend a little on costs c_K/c_T and c_K/c_N, and are determined approximately by μK. ∎

8.2 Periodic Garbage Collection Model

A database is updated and garbage due to update accumulates in the storage area that is the same model as that of Section 8.1. However, the information for the number of updates and the total garbage is collected only at periodic planned times. In this section, the GC is done at periodic times to recover computer resources such as operating time, storage area, and throughput.

It is assumed that a database is updated in a nonhomogeneous Poisson process with an intensity function $h(t)$ that is increasing in t and a mean value function $H(t)$. Introducing the mean times of GC that depend on the number of updates and amount of garbage, the availabilities are obtained, and optimum times T^* that minimize them are discussed analytically.

(1) Model 1 with Number of Updates

Suppose that an amount of garbage arises from the jth $(j = 1, 2, \cdots)$ update with constant probability α $(0 < \alpha \le 1)$ and the mean time required for the collection of this garbage is $c_0(j)$ that depends only on the number of updates, where $c_0(0) \equiv 0$. The mean time for GC at time T is c_T when the total number of updates is less than a prespecified N and is c_N when it is equal to N or has exceeded N until time T. It is assumed that $c_0(j)$ is increasing in j and $c_T \le c_N$. Under these conditions, the mean time for GC at time T is

$$\sum_{j=0}^{N-1} p_j(T) \left[c_T + \sum_{i=0}^{j} \alpha c_0(i) \right] + \sum_{j=N}^{\infty} p_j(T) \left[c_N + \sum_{i=0}^{j} \alpha c_0(i) \right]$$

$$= c_N - (c_N - c_T) \sum_{j=0}^{N-1} p_j(T) + \sum_{j=0}^{\infty} p_j(T) \sum_{i=0}^{j} \alpha c_0(i), \qquad (8.15)$$

where $p_j(t) \equiv \{[H(t)]^j/j!\} e^{-H(t)}$ $(j = 0, 1, 2, \cdots)$.

Suppose that a database can be updated at every time T, although processing efficiency may be worse when the total number of updates has exceeded N. Then, the availability is, from (3.10),

$$A_1(T) = \frac{T}{T + c_N - (c_N - c_T)\sum_{j=0}^{N-1} p_j(T) + \sum_{j=0}^{\infty} p_j(T)\sum_{i=0}^{j} \alpha c_0(i)}.$$

(8.16)

We seek an optimum GC time T_1^* that maximizes $A_1(T)$ in (8.16). Differentiating $A_1(T)$ with respect to T and setting it equal to zero,

$$(c_N - c_T)\left[Th(T)p_{N-1}(T) + \sum_{j=0}^{N-1} p_j(T)\right]$$

$$+ Th(T)\sum_{j=0}^{\infty} p_j(T)\alpha c_0(j+1) - \sum_{j=0}^{\infty} p_j(T)\sum_{i=0}^{j}\alpha c_0(i) = c_N.$$

(8.17)

First, consider the particular case of $c_N = c_T$. Then, (8.17) is

$$Th(T)\sum_{j=0}^{\infty} p_j(T)\alpha c_0(j+1) - \sum_{j=0}^{\infty} p_j(T)\sum_{i=0}^{j}\alpha c_0(i) = c_N.$$

(8.18)

It is assumed that either $h(t)$ or $c_0(j)$ is strictly increasing. Letting $Q(T)$ be the left-hand side of (8.18), $Q(0) = 0$ and

$$\frac{dQ(T)}{dT} = T\left\{\frac{dh(T)}{dT}\sum_{j=0}^{\infty} p_j(T)\alpha c_0(j+1)\right.$$

$$\left. + [h(T)]^2\sum_{j=0}^{\infty} p_j(T)\alpha[c_0(j+2) - c_0(j+1)]\right\} > 0.$$

Thus, if $Q(\infty) \equiv \lim_{T\to\infty} Q(T) > c_N$, then there exists a finite and unique T_0^* that satisfies (8.18). If $h(t)$ is strictly increasing, we easily find that, for any $T > T_0$,

$$Q(T) > h(T)T_0\sum_{j=0}^{\infty} p_j(T_0)\alpha c_0(j+1) - \sum_{j=0}^{\infty} p_j(T_0)\sum_{i=0}^{j}\alpha c_0(i).$$

Hence, if $h(t)$ is strictly increasing to infinity, then a finite T_0^* exists uniquely. When $c_0(j)$ is constant, i.e., $c_0(j) \equiv c_0$, (8.18) is

$$Th(T) - H(T) = \frac{c_N}{\alpha c_0},$$

(8.19)

that agrees with (4.18) of [1] in the periodic replacement with minimal repair at failure. Thus, if a solution T_1^* to (8.19) exists, then it is unique.

Furthermore, when a database is updated in a Poisson process, i.e., $h(t) = \lambda$ and $p_j(t) = [(\lambda t)^j/j!]e^{-\lambda t}$, the left-hand side of (8.18) is

$$\lambda T \sum_{j=0}^{\infty} \alpha c_0(j+1) p_j(T) - \sum_{j=0}^{\infty} \alpha c_0(j+1) \int_0^T \lambda p_j(t) \, dt$$

$$= \lambda \sum_{j=0}^{\infty} \alpha [c_0(j+2) - c_0(j+1)] \int_0^T (\lambda t) p_j(t) \, dt. \tag{8.20}$$

Thus, if $c_0(j)$ is strictly increasing in j, then (8.20) is also strictly increasing in T from 0 to $\alpha \sum_{j=1}^{\infty} [c_0(\infty) - c_0(j)]$, where $c_0(\infty) \equiv \lim_{j \to \infty} c_0(j)$. Hence, if $\alpha \sum_{j=1}^{\infty} [c_0(\infty) - c_0(j)] > c_N$, a finite T_0^* exists uniquely.

Therefore, because the left-hand side of (8.17) is greater than $Q(T)$ for $c_N > c_T$, if either $h(t)$ or $c_0(j)$ is strictly increasing and $Q(\infty) > c_N$, then $T_0^* \geq T_1^*$.

Next, suppose that a database becomes impossible for any updates and the GC is done immediately when the total number of updates has exceeded N before time T. Then, the mean time to GC is

$$T \sum_{j=0}^{N-1} p_j(T) + \int_0^T t \, h(t) p_{N-1}(t) \, dt = \sum_{j=0}^{N-1} \int_0^T p_j(t) \, dt,$$

and by a similar method for obtaining (8.15), the mean time for GC is

$$\sum_{j=0}^{N-1} p_j(T) \left[c_T + \sum_{i=0}^{j} \alpha c_0(i) \right] + \sum_{j=N}^{\infty} p_j(T) \left[c_N + \sum_{i=0}^{N} \alpha c_0(i) \right]$$

$$= c_N - (c_N - c_T) \sum_{j=0}^{N-1} p_j(T) + \sum_{i=1}^{N} \alpha c_0(i) \sum_{j=i}^{\infty} p_j(T). \tag{8.21}$$

In this case, the availability is

$$\tilde{A}_1(T) = \frac{\sum_{j=0}^{N-1} \int_0^T p_j(t) \, dt}{\sum_{j=0}^{N-1} \int_0^T p_j(t) \, dt + c_N - (c_N - c_T) \sum_{j=0}^{N-1} p_j(T)}{+ \sum_{i=1}^{N} \alpha c_0(i) \sum_{j=i}^{\infty} p_j(T)}. \tag{8.22}$$

In particular, by setting that $p_0(t) = \overline{F}(t)$ when $N = 1$,

$$\tilde{A}_1(T) = \frac{\int_0^T \overline{F}(t) \, dt}{\int_0^T \overline{F}(t) \, dt + c_T + [c_N - c_T + \alpha c_0(1)] F(T)}, \tag{8.23}$$

that agrees with (6.13) of [1] when $\alpha = 0$. That is, the policy maximizing $\tilde{A}_1(T)$ corresponds to the policy maximizing the availability of a one-unit system with repair and preventive maintenance.

(2) Model 2 with Amount of Garbage

Suppose that an amount of garbage arises from each update according to a probability distribution $G(x)$ and the total garbage is additive. The distribution of the total garbage at the jth update is $G^{(j)}(x)$, where $G^{(j)}(x)$ $(j = 1, 2, \cdots)$ is the j-fold convolution of $G(x)$ and $G^{(0)}(x) \equiv 1$ for $x \geq 0$. Furthermore, the mean time required for the collection of this garbage is $c_0(x)$ that depends only on its amount and increases from $c_0(0) = 0$. The mean time for GC at time T is c_T when the total garbage is less than an upper limit level K and is c_K with $c_K \geq c_T$ when it has exceeded K. Under this policy, the mean time for GC at time T is

$$\sum_{j=0}^{\infty} p_j(T) \int_0^K [c_T + c_0(x)] \, \mathrm{d}G^{(j)}(x) + \sum_{j=0}^{\infty} p_j(T) \int_K^{\infty} [c_K + c_0(x)] \, \mathrm{d}G^{(j)}(x)$$

$$= c_K - (c_K - c_T) \sum_{j=0}^{\infty} p_j(T) G^{(j)}(K) + \sum_{j=0}^{\infty} p_j(T) \int_0^{\infty} c_0(x) \, \mathrm{d}G^{(j)}(x). \quad (8.24)$$

Therefore, the availability is

$$A_2(T) = \frac{T}{\begin{aligned} &T + c_K - (c_K - c_T) \sum_{j=0}^{\infty} p_j(T) G^{(j)}(K) \\ &+ \sum_{j=0}^{\infty} p_j(T) \int_0^{\infty} c_0(x) \mathrm{d}G^{(j)}(x) \end{aligned}}. \quad (8.25)$$

Differentiating $A_2(T)$ with respect to T and setting it equal to zero,

$$(c_K - c_T) \left\{ Th(T) \sum_{j=0}^{\infty} p_j(T)[G^{(j)}(K) - G^{(j+1)}(K)] + \sum_{j=0}^{\infty} p_j(T) G^{(j)}(K) \right\}$$

$$+ Th(T) \sum_{j=0}^{\infty} p_j(T) \int_0^{\infty} c_0(x) \, \mathrm{d}[G^{(j+1)}(x) - G^{(j)}(x)]$$

$$- \sum_{j=0}^{\infty} p_j(T) \int_0^{\infty} c_0(x) \, \mathrm{d}G^{(j)}(x) = c_K. \quad (8.26)$$

We can make discussions similar to those of the case (1).

Suppose that a database becomes impossible for any updates and the GC is done immediately, when the total garbage has exceeded K before time T. Then, the mean time to GC is

$$T \sum_{j=0}^{\infty} G^{(j)}(K) p_j(T) + \sum_{j=0}^{\infty} [G^{(j)}(K) - G^{(j+1)}(K)] \int_0^T t \, p_j(t) h(t) \, \mathrm{d}t$$

$$= \sum_{j=0}^{\infty} G^{(j)}(K) \int_0^T p_j(t) \, \mathrm{d}t, \quad (8.27)$$

and the mean time for GC is

$$\sum_{j=0}^{\infty} p_j(T) \int_0^K [c_T + c_0(x)] \, dG^{(j)}(x)$$

$$+ \sum_{j=0}^{\infty} \int_0^T p_j(t) h(t) \, dt \int_0^K \left\{ \int_{K-y}^{\infty} [c_K + c_0(x+y)] \, dG(x) \right\} dG^{(j)}(y). \quad (8.28)$$

Therefore, the availability is

$$\tilde{A}_2(T) = \frac{\sum_{j=0}^{\infty} G^{(j)}(K) \int_0^T p_j(t) \, dt}{\begin{array}{l} \sum_{j=0}^{\infty} G^{(j)}(K) \int_0^T p_j(t) \, dt + c_K - (c_K - c_T) \sum_{j=0}^{\infty} p_j(T) G^{(j)}(K) \\ + \sum_{j=0}^{\infty} p_j(T) \int_0^K c_0(x) \, dG^{(j)}(x) \\ + \sum_{j=0}^{\infty} \int_0^T p_j(t) h(t) \, dt \int_0^K [\int_{K-y}^{\infty} c_0(x+y) \, dG(x)] \, dG^{(j)}(y) \end{array}}.$$

$$(8.29)$$

Example 8.2.   We compute optimum times T_i^* numerically that maximize $A_i(T)$ $(i = 1, 2)$ in (8.16) and (8.25), respectively, when $h(t) = \lambda$, $G(x) = 1 - e^{-\mu x}$, and $c_0(x) = c_0 x$, i.e., the mean time to collect garbage increases in proportion to the number of updates or the total garbage and $p_j(t) = [(\lambda t)^j/j!] e^{-\lambda t}$. Then, from (8.17), an optimum T_1^* satisfies

$$(c_N - c_T) \left[\lambda T p_{N-1}(T) + \sum_{j=0}^{N-1} p_j(T) \right] + c_0 \alpha \frac{(\lambda T)^2}{2} = c_N.$$

When N goes to infinity, an optimum time is given by

$$\tilde{T}_1 = \frac{1}{\lambda} \sqrt{\frac{2c_T}{\alpha c_0}}.$$

From (8.26), an optimum T_2^* satisfies

$$\lambda T \sum_{j=0}^{\infty} p_j(T) \frac{(\mu K)^j}{j!} e^{-\mu K} - \sum_{j=1}^{\infty} p_j(T) \sum_{i=0}^{j-1} \frac{(\mu K)^i}{i!} e^{-\mu K} = \frac{c_T}{c_K - c_T}.$$

Tables 8.3 and 8.4 present T_1^* and T_2^* for $N = \mu K = 300, 500, 700$, $c_k/c_T = 2, 5, 10$ $(k = N, K)$, and $c_T/c_0 = 3, 5, 10, 20$ when $\alpha = 10^{-4}$ and $\lambda = 10$, and $\tilde{T}_1$ when $N = \infty$. Optimum T_1^* are strictly increasing in N to $\tilde{T}_1$. From the assumption of $N = \mu K$, optimum times are almost the same ones. From this example, when $N = \mu K = 500$, $c_k/c_T = 2$, and $c_T/c_0 = 20$, T_1^* and T_2^* are about 44, that is, when a database is updated 10 times an hour and exceeds a limit level at 50 hours, on average, the GC should be done at 44 hours, i.e., at about 5.5 days when it is used for 8 hours a day. This also indicates that $\tilde{T}_1$ when $N = \infty$ is approximately good when N is large and c_T/c_0 is small.

Table 8.3. Optimum time T_1^* when $\alpha = 10^{-4}$, $\lambda = 10$, and $\tilde{T}_1$ when $N = \infty$

N	c_T/c_0	c_N/c_T		
		2	5	10
	3	24.3	24.1	23.9
300	5	25.9	25.2	24.8
	10	26.4	25.5	25.1
	20	26.6	25.7	25.2
	3	24.5	24.5	24.5
500	5	31.6	31.6	31.6
	10	43.3	42.7	42.3
	20	44.8	43.8	43.3
	3	24.5	24.5	24.5
700	5	31.6	31.6	31.6
	10	44.7	44.7	44.7
	20	61.7	61.0	60.6
	3		24.5	
∞	5		31.6	
	10		44.7	
	20		63.2	

Next, when $h(t) = \lambda$ and $c_0(j) = c_0$ for Model 1, a finite T_1^* does not exist. However, there exists a finite and unique $\tilde{T}_1^*$ to maximize $\tilde{A}_1(T)$ in (8.22) for $c_N > c_T$ that satisfies

$$\frac{\lambda p_{N-1}(T) \sum_{j=0}^{N-1} \int_0^T p_j(t)\,dt}{\sum_{j=0}^{N-1} p_j(T)} + \sum_{j=0}^{N-1} p_j(T) = \frac{c_N}{c_N - c_T}.$$

Table 8.5 indicates the optimum time $\tilde{T}_1^*$ for $N = 300, 500, 700$ and $c_N/c_T = 2, 5, 10$. These optimum values are $\tilde{T}_1^* > T_1^*$, however, almost the same as those in Table 8.3 when $c_T/c_0 = 20$. ∎

If a database is updated in a Poisson process and the mean time to collect garbage is constant, then the latter modified model of Model 1 would be more practical than the first one. Moreover, by modifying these models, we would consider some models where the GC should be done at the number of updates, the amount of garbage, or the memory areas.

We have assumed until now that c_k $(k = T, N, K)$ represents as the time for the GC at k. If c_k is denoted as the cost for the GC at k, the availabilities derived in the section can be easily converted to the expected cost rates as follows: The expected cost rates of Model 1 are, from (8.16) and (8.22), respectively,

$$C_1(T) = \frac{1}{T}\left[c_N - (c_N - c_T)\sum_{j=0}^{N-1} p_j(T) + \sum_{j=0}^{\infty} p_j(T) \sum_{i=0}^{j} \alpha c_0(i)\right], \quad (8.30)$$

Table 8.4. Optimum time T_2^* when $\lambda = 10$

μK	c_K/c_T		
	2	5	10
300	26.2	24.5	23.8
500	44.3	42.5	41.6
700	62.9	60.8	59.7

Table 8.5. Optimum time $\widetilde{T}_1^*$ when $\lambda = 10$

N	c_N/c_T		
	2	5	10
300	26.7	25.8	25.3
500	45.5	44.3	43.7
700	64.4	63.0	62.3

and

$$\widetilde{C}_1(T) = \frac{c_N - (c_N - c_T)\sum_{j=0}^{N-1} p_j(T) + \sum_{i=1}^{N} \alpha c_0(i) \sum_{j=i}^{\infty} p_j(T)}{\sum_{j=0}^{N-1} \int_0^T p_j(t)\,\mathrm{d}t}. \tag{8.31}$$

The expected cost rates of Model 2 are, from (8.25) and (8.29), respectively,

$$C_2(T) = \frac{1}{T}\left[c_K - (c_K - c_T)\sum_{j=0}^{\infty} p_j(T)G^{(j)}(K) + \sum_{j=0}^{\infty} p_j(T)\int_0^{\infty} c_0(x)\,\mathrm{d}G^{(j)}(x) \right], \tag{8.32}$$

and

$$\widetilde{C}_2(T) = \frac{\begin{array}{l} c_K - (c_K - c_T)\sum_{j=0}^{\infty} p_j(T)G^{(j)}(K) \\ + \sum_{j=0}^{\infty} p_j(T)\int_0^K c_0(x)\,\mathrm{d}G^{(j)}(x)] \\ + \sum_{j=0}^{\infty} \int_0^T p_j(t)h(t)\,\mathrm{d}t\int_0^K [\int_{K-y}^{\infty} c_0(x+y)\,\mathrm{d}G(x)]\,\mathrm{d}G^{(j)}(y) \end{array}}{\sum_{j=0}^{\infty} G^{(j)}(K)\int_0^T p_j(t)\,\mathrm{d}t}. \tag{8.33}$$

8.3 Modified Periodic Garbage Collection Model

We apply the condition-based preventive maintenance in Section 6.1 to the GC model with an upper limit level K of the total garbage: A database is updated in a nonhomogeneous Poisson process with a mean value function $H(t)$. An amount W_j of garbage arises from the jth update and has a probability distribution $G(x) \equiv \Pr\{W_j \le x\}$ $(j = 1, 2, \cdots)$, and the garbage is additive. The total garbage is checked at periodic times nT $(n = 1, 2, \cdots)$, i.e., it is investigated only through checking of space areas and storage conditions in the database. Any maintenance is not done if the total garbage is less than a managerial level Z $(0 \leq Z \leq K)$. On the other hand, if the total garbage has exceeded Z during $(nT, (n+1)T]$, the GC is done at time $(n+1)T$ and the database is restored to its original state.

Let c_K be a loss cost for a useless database when the total garbage is equal to K, and c_Z be a loss cost for the GC where $c_Z < c_K$ when the total garbage has exceeded Z. Then, from (6.4), the expected cost rate for the GC policy is

$$
C(Z) = \frac{\begin{aligned}&c_Z + (c_K - c_Z) \sum_{n=0}^{\infty} \sum_{j=0}^{\infty} p_j[H(nT)] \\ &\times \sum_{i=0}^{\infty} p_i[H((n+1)T) - H(nT)] \int_0^Z [1 - G^{(i)}(K-x)] \, dG^{(j)}(x)\end{aligned}}{\begin{aligned}&\sum_{n=0}^{\infty} \sum_{j=0}^{\infty} p_j[H(nT)] \sum_{i=0}^{\infty} \int_0^Z G^{(i)}(K-x)] \, dG^{(j)}(x) \\ &\times \int_{nT}^{(n+1)T} p_i[H(t) - H(nT)] \, dt\end{aligned}}.
$$

(8.34)

In particular, when a database is updated in a Poisson process, i.e., $H(t) = \lambda t$, the expected cost rate is rewritten as

$$
C(Z) = \frac{\begin{aligned}&c_Z + (c_K - c_Z) \sum_{n=0}^{\infty} \sum_{j=0}^{\infty} p_j(n\lambda T) \\ &\times \sum_{i=0}^{\infty} p_i(\lambda T) \int_0^Z [1 - G^{(i)}(K-x)] \, dG^{(j)}(x)\end{aligned}}{\sum_{n=0}^{\infty} \sum_{j=0}^{\infty} p_j(n\lambda T) \sum_{i=0}^{\infty} \int_0^Z G^{(i)}(K-x)] \, dG^{(j)}(x) \int_0^T p_i(\lambda t) \, dt},
$$

(8.35)

where $p_j(t) \equiv [(\lambda t)^j / j!] e^{-\lambda t}$ $(j = 0, 1, 2, \cdots)$. The optimum GC policy from Section 6.1.2 is given as follows:

(i) If $M_G(K) > c_Z/(c_K - c_Z)$, then there exists a unique Z^* $(0 < Z^* < K)$ that satisfies

$$
Q(Z) \sum_{n=0}^{\infty} \sum_{j=0}^{\infty} p_j(n\lambda T) \sum_{i=0}^{\infty} \int_0^T p_i(\lambda t) \, dt \int_0^Z G^{(i)}(K-x)] \, dG^{(j)}(x)
$$
$$
- \sum_{n=0}^{\infty} \sum_{j=0}^{\infty} p_j(n\lambda T) \sum_{i=0}^{\infty} p_i(\lambda T) \int_0^Z [1 - G^{(i)}(K-x)] \, dG^{(j)}(x) = \frac{c_Z}{c_K - c_Z},
$$

(8.36)

where $M_G(K) \equiv \sum_{j=1}^{\infty} G^{(j)}(K)$ and

$$
Q(Z) \equiv \frac{\sum_{i=0}^{\infty} p_i(\lambda T)[1 - G^{(i)}(K-Z)]}{\sum_{i=0}^{\infty} \int_0^T p_i(\lambda t) \, dt G^{(i)}(K-Z)} \qquad (0 \leq Z \leq K).
$$

In this case, the expected cost rate is

$$
C(Z^*) = (c_K - c_Z) Q(Z^*).
$$

(8.37)

(ii) If $M_G(K) \leq c_Z/(c_K - c_Z)$, then $Z^* = K$, i.e., the GC is done after the total garbage becomes K, and the resulting cost rate is given in (3.12).

Table 8.6. Optimum garbage rate Z^*/K to minimize $C(Z)$

λT	μK	c_K/c_Z			
		100	200	500	1000
60	300	0.708	0.696	0.683	0.673
	500	0.825	0.818	0.810	0.804
	700	0.875	0.870	0.864	0.860
	1000	0.912	0.909	0.905	0.902
120	300	0.963	0.923	0.802	0.702
	500	0.978	0.954	0.881	0.821
	700	0.984	0.976	0.915	0.872
	1000	0.989	0.977	0.941	0.911

Example 8.3. We compute the optimum policy numerically when $G(x) = 1 - e^{-\mu x}$ and $\widetilde{p}(x) = [(\mu x)^j/j!]e^{-\mu x}$ $(j = 0, 1, 2, \dots)$. In this case, (8.36) is

$$
Q(Z) \sum_{n=0}^{\infty} \sum_{j=0}^{\infty} p_j(n\lambda T) \sum_{i=0}^{\infty} \int_0^T p_i(\lambda t)\, \mathrm{d}t
$$
$$
\times \left[1 - \sum_{k=0}^{j-1} \widetilde{p}_k(\mu Z) - \sum_{k=0}^{i-1} \sum_{l=0}^{k} \widetilde{p}_{k-l}(\mu(K - Z))\widetilde{p}_{l+j}(\mu Z) \right]
$$
$$
- \sum_{n=0}^{\infty} \sum_{j=0}^{\infty} p_j(n\lambda T) \sum_{i=0}^{\infty} p_i(\lambda T) \sum_{k=0}^{i-1} \sum_{l=0}^{k} \widetilde{p}_{k-l}(\mu(K - Z))\widetilde{p}_{l+j}(\mu Z)
$$
$$
= \frac{c_Z}{c_K - c_Z}, \tag{8.38}
$$

where $\sum_0^{-1} \equiv 0$. From optimum policy (i), if $\mu K > c_Z/(c_K - c_Z)$, then a finite Z^* to satisfy (8.38) exists uniquely.

Suppose that a database is updated in a Poisson process and the expected number of updates during any interval $(nT, (n+1)T]$ is $H((n+1)T) - H(nT) = \lambda T = 60, 120$. An upper limit level of the total garbage is $\mu K = 300, 500, 700, 1000$. For example, when $\mu K = 700$, the database becomes useless at 700 updates, on average. In addition, when $\lambda T = 120$ and $\lambda = 5$, the expected number of updates is 120 times a day, and hence, the database becomes useless at $700/120 \approx 5.8$ days.

Under the above conditions, Table 8.6 presents the optimum garbage rate Z^*/K for an upper limit level when $c_K/c_Z = 100, 200, 500, 1000$. This example indicates that the optimum value Z^* to minimize the expected cost rate increases with K and decreases with cost rate c_K/c_Z. For example, when $\lambda T = 120$, $\mu K = 700$, and $c_K/c_Z = 1000$, the optimum value is $Z^*/K = 0.872$. If the total garbage has exceeded 87.2% of an upper limit level K, then the GC is done. In this case, the expected number of updates is about $700 \times 0.872 \approx 610$ times. Hence, if $\lambda = 5$, then it is the most economical that the GC is done at the interval $610/120 \approx 5$ days. ∎

9

Backup Policies for a Database System

In recent years, a database in computers systems has become of great importance in modern society with high information. In particular, a reliable database is the most indispensable instrument in on-line transaction processing systems such as real-time systems used for bank accounts. For instance, some errors in the on-line system of a bank might cause social confusion even for a short time, and occasionally, a bank might lose valuable public confidence with oneself.

The data in a computer system are frequently updated by adding or deleting them, and are stored in secondary media. However, data files in secondary media are sometimes broken by several errors due to noise, human errors, and hardware faults. In this case, we have to reconstruct the same files from the beginning. The most simple and dependable method to ensure the safety of data would be always to score the backup copies of all files in other places, and to take them out if some files in the original secondary media are broken. This is called a total backup. But, this method would take hours and be costly when files become very large. To make the backup copies efficiently, we might dump only files that have changed since the last backup. This would reduce significantly both the duration time and the backup size [228]. This is called an export backup.

The total backup is a physical backup scheme that copies all files from the original secondary media into other places. On the other hand, the export backup is a logical backup scheme that copies the data and the definition of a database, where they are stored in the operating system of binary notation. This is generally classified into three schemes: *incremental backup, cumulative backup*, and *full backup* or *complete backup* [229].

The full backup exports all files, and a database system returns to its initial state by this backup. When the full backup copies are repeated frequently, all images of a database can be secured, however, its operating cost and time are remarkably increased. Thus, the scheme of incremental or cumulative backup is usually adopted, and is suitably executed between the operations of full backups in most database systems. The incremental backup exports only files

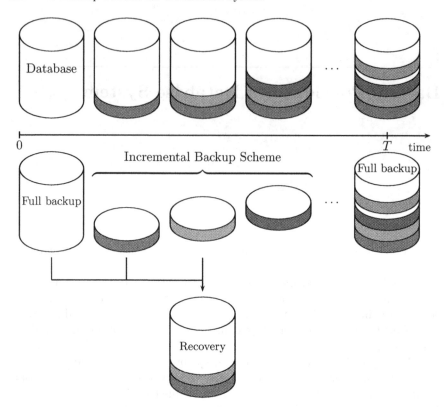

Fig. 9.1. Incremental backup scheme

that have changed since the last incremental or full backup and imports files
of all incremental and the last full backup when some errors have occurred
in storage media (Figure 9.1). Similarly, the cumulative backup exports only
files that have changed since the last full backup and imports files of the
last cumulative and full backups when some errors have occurred. The full
backup with large overhead is done at long intervals and the incremental or
cumulative backup with small overhead is done at short intervals (Figure 9.2).
This could reduce significantly both the duration and cost of backups.

An important problem in actual backup schemes is when to create the full
backup. We want to lessen the number of full backups with large overhead.
However, both overheads of cumulative backup and recovery of incremental
backup increase adaptively with the amount of newly updated trucks. From
this point of view, we have to decide the full backup interval by comparing
two overheads of backup and recovery.

Some recovery techniques for database failures were taken up [230, 231].
Optimum checkpoint intervals of such models that minimize the total overhead
were studied [232–235].

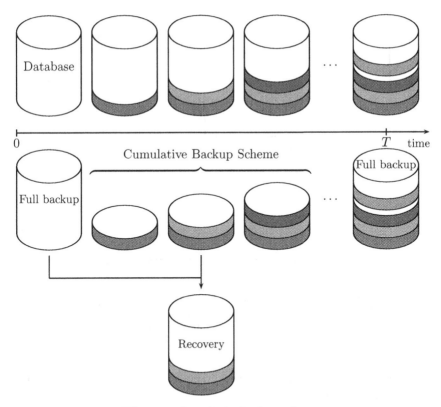

Fig. 9.2. Cumulative backup scheme

In this chapter, we apply the cumulative damage model to the backup of files for database media failures by transforming *shock* into *update* and *damage* into *dumped files* [59, 236, 237].

9.1 Incremental Backup Policy

First, this section considers a modified cumulative damage model with minimal maintenance at shocks in Section 5.4: Suppose that shocks occur in a nonhomogeneous Poisson process and the total damage due to shocks is additive. However, when the total damage has exceeded a threshold level K, it is not additive, and hence, its level is constant at K and minimal maintenance is done at each shock. The damage level remains unchanged by any minimal maintenance. To lessen the maintenance costs after the total damage has exceeded K, the preventive maintenance (PM) is done at a planned time T. The expected cost rate is obtained, and an optimum PM time T^* that minimizes

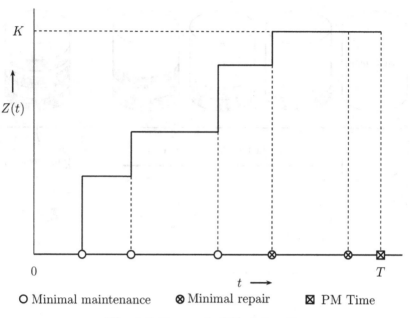

Fig. 9.3. Process for PM at time T

it is discussed analytically in the special case where the times between shocks have an exponential distribution.

Secondly, this model is applied to the backup policy for a database system with secondary storage files when the incremental backup is adopted. Optimum full backup times are computed numerically for several cases.

9.1.1 Cumulative Damage Model with Minimal Maintenance

Consider the cumulative damage model where successive shocks occur at time interval X_j and each shock causes some damage in the amount W_j ($j = 1, 2, \cdots$). It is assumed that $F(t) \equiv \Pr\{X_j \leq t\}$ with finite mean $1/\lambda \equiv \int_0^\infty [1 - F(t)]dt$, and $G_j(x) \equiv \Pr\{W_j \leq x\}$ with finite mean $1/\mu_j \equiv \int_0^\infty [1 - G_j(x)]dx$ ($j = 1, 2, \cdots$).

Suppose that the total damage due to shocks is additive when it has not exceeded a threshold level K, and conversely, it is not additive at any shock after it has exceeded K (Figure 9.3). In this case, the minimal maintenance is done at each shock and the damage level remains in K. Then, the total damage $Z_j \equiv \sum_{i=1}^j W_i$ to the jth shock, where $Z_0 \equiv 0$, has a probability distribution

$$G^{(j)}(x) \equiv \Pr\{Z_j \leq x\} = \begin{cases} 1 & (j = 0), \\ G_1(x) * G_2(x) * \cdots * G_j(x) & (j = 1, 2, \cdots), \end{cases}$$
$$(9.1)$$

where the asterisk mark represents the Stieltjes convolution, $i.e.$, $a(t) * b(t) \equiv \int_0^t b(t-u) da(u)$ for any function $a(t)$ and $b(t)$.

The distribution of the total damage $Z(t)$ defined in (2.1) is, from (2.3),

$$\Pr\{Z(t) \le x\} = \begin{cases} \sum_{j=0}^{\infty} G^{(j)}(x)[F^{(j)}(t) - F^{(j+1)}(t)] & (x \le K), \\ 1 & (x > K), \end{cases} \quad (9.2)$$

and the survival probability is

$$\Pr\{Z(t) > x\} = \begin{cases} \sum_{j=0}^{\infty}[G^{(j)}(x) - G^{(j+1)}(x)]F^{(j+1)}(t) & (x \le K), \\ 0 & (x > K), \end{cases} \quad (9.3)$$

where $F^{(j)}(t)$ ($j = 1, 2, \cdots$) is the j-fold Stieltjes convolution of $F(t)$ and $F^{(0)}(t) \equiv 1$ for $t \ge 0$. Thus, the total expected damage at time t is given by

$$E\{Z(t)\} = \sum_{j=1}^{\infty}[F^{(j)}(t) - F^{(j+1)}(t)] \int_0^K [1 - G^{(j)}(x)]\, dx. \quad (9.4)$$

Suppose that the minimal maintenance for the above model is done at each shock and the damage level remains unchanged by any minimal maintenance. To lessen the maintenance costs after the total damage has exceeded K, the PM is done at a planned time T ($0 < T \le \infty$). The expected number of minimal maintenance, $i.e.$, the expected number of shocks in $[0, T]$ before the total damage has exceeded K is

$$\sum_{j=1}^{\infty} j[F^{(j)}(T) - F^{(j+1)}(T)]G^{(j)}(K). \quad (9.5)$$

Furthermore, the expected number of minimal maintenance actions in $[0, T]$ in the case where the total damage remains in K when it has reached K is

$$\sum_{j=0}^{\infty}[G^{(j)}(K) - G^{(j+1)}(K)]$$

$$\times \sum_{i=0}^{\infty}(i+1) \int_0^T [F^{(i)}(T-t) - F^{(i+1)}(T-t)]\, dF^{(j+1)}(t)$$

$$= \sum_{j=1}^{\infty} F^{(j)}(T)[1 - G^{(j)}(K)], \quad (9.6)$$

and the expected number of minimal maintenance actions in $[0, T]$ in the case where the total damage is less than K when it has reached K is

$$\sum_{j=1}^{\infty} j F^{(j+1)}(T)[G^{(j)}(K) - G^{(j+1)}(K)]. \quad (9.7)$$

Thus, the total expected number of minimal maintenance actions in $[0,T]$ in the case where the total damage is less than K is the sum of (9.5) and (9.7) and is given by

$$\sum_{j=1}^{\infty} F^{(j)}(T)G^{(j)}(K). \tag{9.8}$$

It is evident that $(9.6) + (9.8) = \sum_{j=1}^{\infty} F^{(j)}(T) \equiv M_F(T)$ that represents the expected number of shocks in $[0,T]$.

(1) Expected Cost

We introduce the following costs: The PM cost at time T is $c_K + c_0(K)$ when the total damage has reached a threshold level K, and $c_K + c_0(x)$ when the total damage is x $(0 \le x \le K)$. Then, from (9.3), the PM cost when the total damage is K is

$$[c_K + c_0(K)] \sum_{j=0}^{\infty} F^{(j+1)}(T)[G^{(j)}(K) - G^{(j+1)}(K)], \tag{9.9}$$

and from (9.2), the PM cost when the total damage is less than K is

$$\sum_{j=0}^{\infty} [F^{(j)}(T) - F^{(j+1)}(T)] \int_0^K [c_K + c_0(x)] \, \mathrm{d}G^{(j)}(x). \tag{9.10}$$

Let c_m and c_M $(c_m < c_M)$ be the respective costs of minimal maintenance at each shock when the total damage is less than K and is K. Then, the expected cost rate is, from (9.6), (9.8), (9.9), and (9.10),

$$C(T) = \frac{1}{T} \Bigg\{ [c_K + c_0(K)] \sum_{j=0}^{\infty} F^{(j+1)}(T)[G^{(j)}(K) - G^{(j+1)}(K)]$$

$$+ \sum_{j=0}^{\infty} [F^{(j)}(T) - F^{(j+1)}(T)] \int_0^K [c_K + c_0(x)] \, \mathrm{d}G^{(j)}(x)$$

$$+ c_M \sum_{j=1}^{\infty} F^{(j)}(T)[1 - G^{(j)}(K)] + c_m \sum_{j=1}^{\infty} F^{(j)}(T)G^{(j)}(K) \Bigg\}. \tag{9.11}$$

If shocks occur in a nonhomogeneous Poisson process with a mean value function $H(t)$, the expected cost rate in (9.11) is rewritten as, replacing $F^{(j)}(t) - F^{(j+1)}(t)$ with $p_j(t) \equiv \{[H(t)]^j/j!\} \, \mathrm{e}^{-H(t)}$,

$$\tilde{C}(T) = \frac{1}{T}\left\{[c_K + c_0(K)]\sum_{j=0}^{\infty}p_j(T)[1 - G^{(j)}(K)]\right.$$

$$+ \sum_{j=0}^{\infty}p_j(T)\int_0^K [c_K + c_0(x)]\,\mathrm{d}G^{(j)}(x)$$

$$\left. + c_M\sum_{j=1}^{\infty}p_j(T)\sum_{i=1}^{j}[1 - G^{(i)}(K)] + c_m\sum_{j=1}^{\infty}p_j(T)\sum_{i=1}^{j}G^{(i)}(K)\right\}.$$

$$(9.12)$$

(2) Optimum Policy

Suppose that shocks occur in a Poisson process with rate λ, i.e., $F(t) = 1 - e^{-\lambda t}$ and $p_j(t) = [(\lambda t)^j/j!]e^{-\lambda t}$ $(j = 0, 1, 2, \cdots)$. In addition, it is assumed that $c_0(x) = c_0 x$, i.e., the PM cost is proportional to the total damage. Then, (9.11) or (9.12) is simplified as

$$C(T) = \frac{1}{T}\left\{c_0\sum_{j=1}^{\infty}p_j(T)\int_0^K [1 - G^{(j)}(x)]\,\mathrm{d}x\right.$$

$$\left. - (c_M - c_m)\sum_{j=1}^{\infty}p_j(T)\sum_{i=1}^{j}G^{(i)}(K) + c_K + c_M\lambda T\right\}. \quad (9.13)$$

We seek an optimum PM time T^* that minimizes $C(T)$ in (9.13). It is clear that $\lim_{T\to 0}C(T) = \infty$ and $\lim_{T\to\infty}C(T) = \lambda c_M$. Thus, there exists a positive T^* $(0 < T^* \leq \infty)$ that minimizes $C(T)$. Differentiating $C(T)$ with respect to T and setting it equal to zero,

$$c_0\left\{\lambda T\sum_{j=0}^{\infty}p_j(T)\int_0^K [G^{(j)}(x) - G^{(j+1)}(x)]\,\mathrm{d}x\right.$$

$$\left. - \sum_{j=1}^{\infty}p_j(T)\int_0^K [1 - G^{(j)}(x)]\,\mathrm{d}x\right\}$$

$$+ (c_M - c_m)\sum_{j=1}^{\infty}p_j(T)\sum_{i=1}^{j}[G^{(i)}(K) - G^{(j)}(K)] = c_K. \quad (9.14)$$

In the particular case of $c_0 = 0$, (9.14) becomes

$$\sum_{j=1}^{\infty}p_j(T)\sum_{i=1}^{j}[G^{(i)}(K) - G^{(j)}(K)] = \frac{c_K}{c_M - c_m}. \quad (9.15)$$

Letting the left-hand side of (9.15) be denoted by $Q(T)$, $\lim_{T\to 0}Q(T) = 0$, $\lim_{T\to\infty}Q(T) = \sum_{j=1}^{\infty}G^{(j)}(K) \equiv M_G(K)$, and

$$\frac{\mathrm{d}Q(T)}{\mathrm{d}T} = \lambda \sum_{j=1}^{\infty} p_j(T)[G^{(j)}(K) - G^{(j+1)}(K)] > 0.$$

Thus, $Q(T)$ is strictly increasing from 0 to $M_G(K)$ that is the expected number of shocks before the total damage exceeds a threshold level K. In this case, we have the following optimum policy:

(i) If $M_G(K) > c_K/(c_M - c_m)$, then there exists a finite and unique T^* $(0 < T^* < \infty)$ that satisfies (9.15).
(ii) If $M_G(K) \leq c_K/(c_M - c_m)$, then $T^* = \infty$, i.e., the PM should not be done.

Note that an optimum T^* $(0 < T^* < \infty)$ always exists for $c_0 > 0$ because the left-hand side of (9.14) increases from 0 to ∞, as $T \to \infty$.

9.1.2 Incremental Backup

We apply the cumulative damage model discussed in Section 9.1.1 to the backup of secondary storage files in a database system. Suppose that a database is updated in a Poisson process with rate λ. To ensure the safety of data and to save costs or hours, we make the following backup policy: When the total dumped files do not exceed a threshold level K, we perform the incremental backup of only new files since the previous backup. Conversely, when the total files have exceeded K, we perform the total backup instead where both the time and size of the backup are constant. In addition, we perform the full backup at periodic times nT $(n = 1, 2, \cdots)$ where all files are dumped and the system returns to its initial state.

Let us introduce the following costs: Cost $c_K + c_0 x$ is incurred for the full backup when the total files are x $(0 \leq x \leq K)$ at periodic times nT, and cost $c_K + c_0 K$ is incurred for the full backup when the total files have exceeded K. Furthermore, let c_m and c_M $(c_m < c_M)$ be the costs for incremental and total backups, respectively. Under such assumptions, the expected cost rate has been already given in (9.13).

In this section, we consider two cases: (1) Backup files due to each update have an identical probability distribution, and (2) backup files due to each update have different probability distributions that increase at a geometric rate.

(1) Identical Distribution

Suppose that backup files due to each update have an identical exponential distribution $G(x)$, i.e., $G^{(j)}(x) = \sum_{i=j}^{\infty}[(\mu x)^i/i!]e^{-\mu x}$ $(j = 0, 1, 2, \cdots)$. Then, because

$$\int_0^K [G^{(j)}(x) - G^{(j+1)}(x)]\,\mathrm{d}x = \frac{1}{\mu}G^{(j+1)}(K),$$

and

$$\int_0^K [1 - G^{(j)}(x)]\,\mathrm{d}x = \frac{1}{\mu} \sum_{i=1}^j G^{(i)}(K),$$

the expected cost rate $C(T)$ in (9.13) and (9.14) is simplified, respectively, as

$$C(T) = \frac{1}{T}\left[c_K + c_M \lambda T - \left(c_M - c_m - \frac{c_0}{\mu} \right) \sum_{j=1}^{\infty} p_j(T) \sum_{i=1}^j G^{(i)}(K) \right], \quad (9.16)$$

and

$$\left(c_M - c_m - \frac{c_0}{\mu} \right) \sum_{j=1}^{\infty} p_j(T) \sum_{i=1}^j [G^{(i)}(K) - G^{(j)}(K)] = c_K, \quad (9.17)$$

where $p_j(t) = [(\lambda t)^j/j!]e^{-\lambda t}$ $(j = 0, 1, 2, \cdots)$. The left-hand side of (9.17) is a strictly increasing function of T from 0 to $(c_M - c_m - c_0/\mu)\mu K$.

Therefore, if $c_M - c_m - c_0/\mu > c_K/(\mu K)$, then there exists a finite and unique T^* that satisfies (9.17), and the resulting cost rate is

$$\frac{C(T^*)}{\lambda} = c_M - \left(c_M - c_m - \frac{c_0}{\mu} \right) \sum_{j=1}^{\infty} p_j(T^*) G^{(j+1)}(K). \quad (9.18)$$

Conversely, if $c_M - c_m - c_0/\mu \le c_K/(\mu K)$, then $T^* = \infty$ and $C(\infty) = \lambda c_M$.

(2) Different Distribution

First, we show that an amount W_j of files that is dumped at the jth update decreases at a geometric ratio. Suppose that an amount of files at some update is W, the total volume of files is M, and the total files that have been already dumped are A $(0 \le A \le M)$. Then, assume that an amount of newly dumped files is proportional to the vacant space, *i.e.*, $W(M - A)/M$. Letting W_j be newly dumped files at the jth update,

$$W_1 = W,$$

$$W_{j+1} = W\frac{M - \sum_{i=1}^j W_i}{M} \quad (j = 1, 2, \cdots).$$

Solving this equation,

$$W_j = W\left(1 - \frac{W}{M} \right)^{j-1} \quad (j = 1, 2, \cdots). \quad (9.19)$$

We set $W/M \equiv 1 - \alpha$ $(0 \le \alpha < 1)$ that is an amount ratio of dumped files at the first update. Then, $W_j/M = (1 - \alpha)\alpha^{j-1}$ $(j = 1, 2, \cdots)$ that is a geometric distribution with mean $1/(1-\alpha)$. This indicates that an amount of newly dumped files is strictly decreasing and forms a geometric process with W/a^{j-1} $(j = 1, 2, \cdots)$, where $1/a \equiv \alpha$ [250].

Furthermore, it is of interest that the total ratio of dumped files until the jth update is

$$\frac{1}{M} \sum_{i=1}^{j} W_i = 1 - \alpha^j \qquad (j = 1, 2, \cdots),$$

that is equal to the reliability of a parallel system with j units each of whose reliabilities is $1 - \alpha$.

It is usually known that an initial estimated amount of dumped files is about 25% and a threshold level K is 60% of the total volume. In this case, the number of updates where the total files exceed K is given by a minimum value that satisfies $1 - (1 - 0.25)^n \geq 0.6$ and its solution is $n = 4$. Conversely, if the number of updates where the total files exceed 60% is $n = 4$, then the amount rate is given by $1 - \alpha^4 \geq 0.6$ and $1 - \alpha$ is larger than 0.205.

Suppose that an amount W_j of newly dumped files at the jth update has an exponential distribution $G_j(x) = 1 - e^{-\mu_j x}$ $(\mu_1 < \mu_2 < \cdots)$. Then, the distribution of total files until the jth update is easily given by

$$G^{(j)}(x) = 1 - \sum_{l=1}^{j} \left(\prod_{i=1,i\neq l}^{j} \frac{\mu_i}{\mu_i - \mu_l} \right) e^{-\mu_l x} \qquad (j = 1, 2, \cdots), \qquad (9.20)$$

where $\sum_{l=1}^{1} \prod_{i=1,i\neq l}^{1} = 1$. In particular, when W_j increases at a geometric ratio $(0 < \alpha < 1)$, i.e., $W_j = \alpha^{j-1}W$ and $1/\mu_j = \alpha^{j-1}/\mu_1 = \alpha^{j-1}/\mu$,

$$G^{(j)}(x) = 1 - \sum_{l=1}^{j} \left(\prod_{i=1,i\neq l}^{j} \frac{1}{1 - \alpha^{i-l}} \right) e^{-\mu x/\alpha^{l-1}} \qquad (j = 1, 2, \cdots). \qquad (9.21)$$

Thus, substituting $G^{(j)}(x)$ in (9.21) in (9.13) and (9.14), respectively, the expected cost rate is

$$C(T) = \frac{1}{T} \left[c_K + c_M \lambda T - \sum_{j=1}^{\infty} p_j(T) \sum_{i=1}^{j} \left(c_M - c_m - \frac{c_0}{\mu} \alpha^{i-1} \right) G^{(i)}(K) \right],$$
$$(9.22)$$

and (9.14) is

$$\sum_{j=1}^{\infty} p_j(T) \sum_{i=1}^{j} \left[\left(c_M - c_m - \frac{c_0}{\mu} \alpha^{i-1} \right) G^{(i)}(K) - \left(c_M - c_m - \frac{c_0}{\mu} \alpha^{j-1} \right) G^{(j)}(K) \right]$$

$$= c_K. \qquad (9.23)$$

Denoting the left-hand side of (9.23) by $Q_1(T)$, when $M_G(K) \equiv \sum_{j=1}^{\infty} G^{(j)}(K) < \infty$, $Q_1(0) \equiv \lim_{T \to 0} Q_1(T) = 0$, and

$$Q_1(\infty) \equiv \lim_{T \to \infty} Q_1(T) = \sum_{j=1}^{\infty} \left(c_M - c_m - \frac{c_0}{\mu} \alpha^{j-1} \right) G^{(j)}(K).$$

Therefore, if $Q_1(\infty) > c_K$, then there exists a finite T^* $(0 < T^* < \infty)$ that satisfies (9.23), and the resulting cost rate is

$$\frac{C(T^*)}{\lambda} = c_M - \sum_{j=1}^{\infty} \left(c_M - c_m - \frac{c_0}{\mu} \alpha^j \right) p_j(T^*) G^{(j+1)}(K). \qquad (9.24)$$

Example 9.1. First, suppose that W_j has an identical exponential distribution $G_j(x) = 1 - e^{-\mu x}$ $(j = 1, 2, \cdots)$, the total volume of files is 3×10^5 trucks, and a threshold level K is 1.2×10^5 and 1.8×10^5 trucks that correspond to 40% and 60% of the total volume, respectively.

Table 9.1 presents the optimum full backup time λT^* and the resulting cost rate $C(T^*)/\lambda$ for $c_K/(c_M - c_m - c_0/\mu) = 1, 2, 5, 10, 15$ and $\mu K = 12, 18$ when $c_M = C(\infty)/\lambda = 6$ and $c_m + c_0/\mu = 5$. This indicates that the optimum T^* increases with both $c_K/(c_M - c_m - c_0/\mu)$ and μK, and $C(T^*)$ increases with $c_K/(c_M - c_m - c_0/\mu)$, and conversely, decreases with μK. However, they are almost unchanged for $c_K/(c_M - c_m - c_0/\mu)$ and μK.

For example, when the mean time between updates is $1/\lambda = 1$ day, the dumped file is $1/\mu = 10^4$ trucks and $K = 1.2 \times 10^5$ trucks, the optimum full backup time T^* is about 9 days for $c_K/(c_M - c_m - c_0/\mu) = 2$. In this case, $\mu K/\lambda = 12$ days represents the mean time until the total dumped files exceed a threshold level K.

Secondly, suppose that the amount W_j of newly dumped files at the jth update has different exponential distributions $G_j(x) = 1 - e^{-\mu_j x}$ $(j = 1, 2, \cdots)$, and W_j decreases at a geometric ratio α $(0 < \alpha < 1)$, i.e., $W_j = \alpha^{j-1} W$ and $1/\mu_j = \alpha^{j-1}/\mu_1 \equiv \alpha^{j-1}/\mu$. Furthermore, the total volume of files is 5×10^5 trucks, a threshold level K is 4×10^5 trucks that corresponds to 80% of the total volume, and the mean amount of dumped files due to the first update is $1/\mu = 10^5$ trucks that corresponds to 25% of the total volume, i.e., $\mu K = 4$.

Table 9.2 presents the optimum full backup time λT^* for $c_K/(c_M - c_m) = 1, 2, 3, 4, 5, 6$ and $\alpha = 1.00, 0.95, 0.90, 0.85, 0.80, 0.75$ when $(c_0/\mu)/(c_M - c_m) = 0.1$. This indicates that the optimum T^* increases when $c_K/(c_M - c_m)$ increases. For example, when the mean time between updates is $1/\lambda = 1$ day, the mean dumped file is $1/\mu = 10^5$ trucks and $K = 4 \times 10^5$ trucks, the optimum time T^* is about 10 days for $c_K/(c_M - c_m) = 3$ and $\alpha = 0.85$.

This also indicates that λT^* decreases when α increases when a finite optimum time exists. For example, when $\alpha = 0.90$, if $c_K/(c_M - c_m) \geq 5.37 - 0.4 = 4.97$, then a finite T^* does not exist. When $\alpha = 0.80$, $M_G(K) = \infty$, i.e., the total dumped files might not exceed K with a certain probability. In this case, when $c_K/(c_M - c_m) \geq 5$, there does not exist a finite T that satisfies (9.23). When $\alpha = 0.75$, no finite T exists for any $c_K/(c_M - c_m) = 1-6$. ∎

Table 9.1. Optimum full backup time λT^* and expected cost rate $C(T^*)/\lambda$ when $c_M = 6$ and $c_m + c_0/\mu = 5$

$\dfrac{c_K}{c_M - c_m - c_0/\mu}$	$\mu K = 12$		$\mu K = 18$	
	λT^*	$C(T^*)/\lambda$	λT^*	$C(T^*)/\lambda$
1	7.462	5.179	11.001	5.112
2	9.084	5.299	12.767	5.196
5	12.469	5.578	16.069	5.403
10	18.856	5.909	20.372	5.679
15	∞	6.000	25.893	5.898

Table 9.2. Optimum full backup time λT^* when $\mu K = 4$ and $(c_0/\mu)/(c_M - c_m) = 0.1$

α	$c_K/(c_M - c_m)$						$M_G(K)$
	1	2	3	4	5	6	
1.00	3.67	5.53	7.82	∞	∞	∞	4
0.95	4.10	6.17	8.66	∞	∞	∞	4.49
0.90	4.46	6.59	8.88	12.21	∞	∞	5.37
0.85	5.04	7.4 7	10.01	13.25	18.67	∞	15.28
0.80	6.12	9.63	14.32	28.85	∞	∞	∞
0.75	∞	∞	∞	∞	∞	∞	∞

9.2 Incremental and Cumulative Backup Policies

The incremental backup exports only files that have changed or are new since the last incremental backup or full backup. On the other hand, the cumulative backup exports only files that have changed or are new since the last full backup. When some errors have occurred in storage media, we can recover a database system by importing files of all incremental backups and the full backup for the incremental backup scheme and by importing files of the last cumulative and full backups for the cumulative backup scheme. The cumulative backup exports more files than the incremental one at each update, however, it imports less files than the incremental one when we recover a database system.

It is an important problem to determine which backup scheme should be adopted as the backup policy. It is supposed that the full backup is planned at time T or when a database system fails, whichever occurs first. Then, we compare two schemes of incremental and cumulative backups, using the results in Section 9.1. Furthermore, we discuss optimum full backup times for the incremental and cumulative backups and compare them numerically.

9.2.1 Expected Cost Rates

We make the same assumptions as those of Section 9.1.2, $G_j(x) = G(x)$ for all j, and $K = \infty$, *i.e.*, the total dumped files are eternally additive. In addition, a database in secondary media fails according to a general distribution $D(t)$ with finite mean $1/\gamma$. Suppose that the full backup is done at a planned time T $(0 < T \le \infty)$ or when a database fails, whichever occurs first.

Let us introduce the following maintenance costs: Cost c_F is incurred for the full backup, and cost $c_K + c_0 x$ is incurred for the incremental backup when the amount of export files at the backup time is x, and for the cumulative backup when the total amount of export files at the backup time is x. The recovery cost is $c_R + c_0 x$ for the cumulative backup if the database fails when the total amount of import files at the recovery time is x, and is $c_R + c_0 x + j c_N$ for the incremental backup when the number of backups is j.

Let denote by

$$M_j = \int_0^\infty (c_K + c_0 x) \, \mathrm{d}G^{(j)}(x)$$
$$= c_K + \frac{j c_0}{\mu},$$
$$N_j = \int_0^\infty (c_R + c_0 x) \, \mathrm{d}G^{(j)}(x)$$
$$= c_R + \frac{j c_0}{\mu}.$$

Note that $j M_1$ is the expected cost of the incremental backup and $\sum_{i=1}^{j} M_i$ is the expected cost of the cumulative backup at the jth update, and N_j is the expected recovery cost of the cumulative backup, and $N_j + j c_N$ is the expected recovery cost of the incremental backup when j numbers of updates have occurred at the failure of the database.

Therefore, the expected cost until the full backup for the incremental and cumulative backups are, respectively,

$$\widetilde{C}_I(T) = c_F + \overline{D}(T) \sum_{j=0}^\infty [F^{(j)}(T) - F^{(j+1)}(T)](jM_1)$$

$$+ \sum_{j=0}^\infty \int_0^T [F^{(j)}(t) - F^{(j+1)}(t)] \, \mathrm{d}D(t)(jM_1 + N_j + jc_N)$$

$$= c_F + c_R D(T)$$

$$+ \left(c_K + \frac{c_0}{\mu} \right) \int_0^T \overline{D}(t) \, \mathrm{d}M_F(t) + \left(c_N + \frac{c_0}{\mu} \right) \int_0^T M_F(t) \, \mathrm{d}D(t), \quad (9.25)$$

and

$$\tilde{C}_C(T) = c_F + \overline{D}(T) \sum_{j=0}^{\infty} [F^{(j)}(T) - F^{(j+1)}(T)] \sum_{i=1}^{j} M_i$$

$$+ \sum_{j=0}^{\infty} \int_0^T [F^{(j)}(t) - F^{(j+1)}(t)] \, \mathrm{d}D(t) \left(\sum_{i=0}^{j} M_i + N_j \right)$$

$$= c_F + c_R D(T) + c_K \int_0^T \overline{D}(t) \, \mathrm{d}M_F(t)$$

$$+ \frac{c_0}{\mu} \left[\int_0^T M_F(t) \, \mathrm{d}D(t) + \sum_{j=1}^{\infty} j \int_0^T \overline{D}(t) \, \mathrm{d}F^{(j)}(t) \right], \qquad (9.26)$$

where $\sum_{i=1}^{0} \equiv 0$, $\overline{D}(t) \equiv 1 - D(t)$, and $M_F(t) \equiv \sum_{j=1}^{\infty} F^{(j)}(t)$.

To compare the two expected costs, we find the difference between them as follows:

$$\tilde{C}_C(T) - \tilde{C}_I(T) = \frac{c_0}{\mu} \sum_{j=1}^{\infty} j \int_0^T \overline{D}(t) \, \mathrm{d}F^{(j+1)}(t) - c_N \int_0^T M_F(t) \, \mathrm{d}D(t). \quad (9.27)$$

Hence, if

$$\frac{c_0}{\mu} \sum_{j=1}^{\infty} j \int_0^T \overline{D}(t) \, \mathrm{d}F^{(j+1)}(t) > c_N \int_0^T M_F(t) \, \mathrm{d}D(t),$$

then the incremental backup is better than the cumulative one when the full backup is done at time T. The smaller the extra cost c_N required for the incremental backup when the database fails, the more the incremental backup is useful as the backup scheme.

(1) Optimum Full Backup Time for Incremental Backup

Consider the optimum policy for the incremental backup. Because the mean time to the full backup is

$$T\overline{D}(T) + \int_0^T t \, \mathrm{d}D(t) = \int_0^T \overline{D}(t) \, \mathrm{d}t, \qquad (9.28)$$

the expected cost rate is, dividing (9.25) by (9.28),

$$C_I(T) = \frac{c_F + c_R D(T) + (c_K + c_0/\mu) \int_0^T \overline{D}(t) \, \mathrm{d}M_F(t)}{+(c_N + c_0/\mu) \int_0^T M_F(t) \, \mathrm{d}D(t)} {\int_0^T \overline{D}(t) \, \mathrm{d}t}. \qquad (9.29)$$

We find an optimum time T_1^* that minimizes $C_I(T)$ when a database is updated in a Poisson process, i.e., $M_F(t) = \lambda t$. Differentiating $C_I(T)$ with respect to T and setting it equal to zero,

$$c_R \left[r(T) \int_0^T \overline{D}(t) \, dt - D(T) \right] + \lambda \left(c_N + \frac{c_0}{\mu} \right) \int_0^T \overline{D}(t)[Tr(T) - tr(t)] \, dt = c_F,$$

(9.30)

where $r(t) \equiv d(t)/\overline{D}(t)$ and $d(t)$ is a density function of $D(t)$. Let $Q_1(T)$ be the left-hand side of (9.30). Then, if the failure rate $r(t)$ is strictly increasing, $Q_1(T)$ is also strictly increasing from 0 to $Q_1(\infty)$. Thus, if $Q_1(\infty) > c_F$, then there exists a finite and unique T^* that satisfies (9.30). Note that if $r(t)$ is strictly increasing to ∞, then $Q_1(\infty) = \infty$. In this case, the resulting cost rate is

$$C_I(T_1^*) = \lambda \left(c_K + \frac{c_0}{\mu} \right) + \left[c_R + \lambda T_1^* \left(c_N + \frac{c_0}{\mu} \right) \right] r(T_1^*).$$

(9.31)

(2) Optimum Full Backup Time for Cumulative Backup

From (9.26) and (9.28), the expected cost rate for the cumulative backup when a database is updated in a Poisson process with rate λ is

$$C_C(T) = \lambda \left(c_K + \frac{c_0}{\mu} \right) + \frac{c_F + c_R D(T) + (\lambda c_0/\mu)[\int_0^T \lambda t \overline{D}(t) \, dt + \int_0^T t \, dD(t)]}{\int_0^T \overline{D}(t) \, dt}.$$

(9.32)

Thus, differentiating $C_C(T)$ with respect to T and setting it equal to zero,

$$c_R \left[r(T) \int_0^T \overline{D}(t) \, dt - D(T) \right] + \frac{\lambda c_0}{\mu} \int_0^T \overline{D}(t)[\lambda(T-t) + Tr(T) - tr(t)] \, dt = c_F.$$

(9.33)

Hence, if $r(t)$ is strictly increasing, then the left-hand side $Q_2(T)$ of (9.33) is also strictly increasing from 0 to $Q_2(\infty)$. Thus, if $Q_2(\infty) > c_F$, then there exists a finite and unique T_2^* that satisfies (9.33). In this case, the resulting cost rate is

$$C_C(T_2^*) = \lambda \left(c_K + \frac{c_0}{\mu} \right) + \frac{\lambda c_0}{\mu} \left[\lambda T_2^* + T_2^* r(T_2^*) \right] + c_R r(T_2^*).$$

(9.34)

Example 9.2.   Suppose that a database is updated in a Poisson process with rate λ, the backup is done with probability α $(0 < \alpha \le 1)$, and it fails with probability $\beta \equiv 1 - \alpha$ at each update time, i.e., $F^{(j)}(t) - F^{(j+1)}(t) = [(\alpha \lambda t)^j/j!] e^{-\alpha \lambda t}$ $(j = 0, 1, 2, \cdots)$, $M_F(t) = \alpha \lambda t$, and $D(t) = 1 - e^{-\beta \lambda t}$. In this case, (9.27) becomes

$$\widetilde{C}_C(T) - \widetilde{C}_I(T) = \lambda \left(\frac{\alpha c_0}{\mu} - \beta c_N \right) \int_0^T \alpha \lambda t e^{-\beta \lambda t} \, dt.$$

(9.35)

Thus, if $\alpha(c_0/\mu) > \beta c_N$, then the incremental backup is better than the cumulative one, and *vice versa*.

Table 9.3. Optimum full backup time λT_1^* and expected cost rate $C_I(T_1^*)/(\lambda c_0/\mu)$ of the incremental backup for $c_N/(c_0/\mu)$ when $c_F/(c_0/\mu) = 64$, $c_K/(c_0/\mu) = 40$, $c_R/(c_0/\mu) = 100$, and $\alpha = 0.98$.

$c_N/(c_0/\mu)$	λT_1^*	$C_I(T_1^*)/(\lambda c_0/\mu)$
20	18.74	49.89
30	15.25	51.45
40	13.17	52.76
49	11.88	52.82
50	11.76	53.94

First, when the incremental backup is adopted, (9.30) is rewritten as

$$\alpha\lambda T - \frac{\alpha}{\beta}(1 - e^{-\beta\lambda T}) = \frac{c_F}{c_N + c_0/\mu}, \tag{9.36}$$

whose left-hand side is strictly increasing from 0 to ∞. Thus, there exists a finite and unique T_1^* that satisfies (9.36), and the resulting cost rate is

$$\frac{C_I(T_1^*)}{\lambda} = \alpha c_K + \beta c_R + \alpha\beta\lambda T_1^* c_N + (1 + \beta\lambda T_1^*)\frac{\alpha c_0}{\mu}. \tag{9.37}$$

Note from (9.36) that the optimum T_1^* does not depend on c_K and c_R. Table 9.3 presents the optimum full backup time T_1^* and the expected cost rate $C_I(T_1^*)/(\lambda c_0/\mu)$ of the incremental backup for $c_N/(c_0/\mu) = 20, 30, 40, 50$ when $c_F/(c_0/\mu) = 64$, $c_K/(c_0/\mu) = 40$, $c_R/(c_0/\mu) = 100$, and $\alpha = 0.98$. Note that all costs are relative to cost c_0/μ and all times are relative to $1/\lambda$. For example, when $c_N/(c_0/\mu) = 30$, λT_1^* is about 15.25, that is, when the mean time of update is $1/(\alpha\lambda) = 1$ day, the optimum T_1^* is about 15 days.

Secondly, when the cumulative backup is adopted, (9.33) is

$$\alpha\lambda T - \frac{\alpha}{\beta}(1 - e^{-\beta\lambda T}) = \frac{\beta c_F}{c_0/\mu}, \tag{9.38}$$

whose left-hand side is equal to that of (9.36), and the resulting cost rate is

$$\frac{C_C(T_2^*)}{\lambda} = \alpha c_K + \beta c_R + (1 + \lambda T_2^*)\frac{\alpha c_0}{\mu}. \tag{9.39}$$

From the above results, if $c_N/(c_0/\mu) < \alpha/\beta$, then T_1^* is larger than T_2^* and *vice versa*. In this example, when $c_N/(c_0/\mu) = 49$, $\lambda T_1^* = \lambda T_2^* = 11.88$ and $C_I(T_1^*)/(\lambda c_0/\mu) = C_C(T_2^*)/(\lambda c_0/\mu) = 52.82$. Hence, if $c_N/(c_0/\mu) < 49$, then the incremental backup is better than the cumulative one. ∎

9.3 Optimum Full Backup Level for Cumulative Backup

In this section, we derive an optimum full backup level for the cumulative backup. Suppose that we do the full backup when the total files have exceeded a managerial level K ($0 \le K \le \infty$) or when the recovery is completed if the database fails, whichever occurs first. The cumulative backup is done at each update between the full backups.

Underlying the same assumptions as those of Section 9.2, the probability that the full backup is done when the total files have exceeded K is

$$\sum_{j=0}^{\infty} [G^{(j)}(K) - G^{(j+1)}(K)] \int_0^{\infty} \overline{D}(t) \, \mathrm{d}F^{(j+1)}(t), \tag{9.40}$$

and the probability that it is done when the database fails is

$$\sum_{j=0}^{\infty} G^{(j)}(K) \int_0^{\infty} [F^{(j)}(t) - F^{(j+1)}(t)] \, \mathrm{d}D(t), \tag{9.41}$$

where $(9.40) + (9.41) = 1$. Furthermore, the mean time to the full backup is

$$\sum_{j=0}^{\infty} [G^{(j)}(K) - G^{(j+1)}(K)] \int_0^{\infty} t \, \overline{D}(t) \, \mathrm{d}F^{(j+1)}(t)$$

$$+ \sum_{j=0}^{\infty} G^{(j)}(K) \int_0^{\infty} t \, [F^{(j)}(t) - F^{(j+1)}(t)] \, \mathrm{d}D(t)$$

$$= \sum_{j=0}^{\infty} G^{(j)}(K) \int_0^{\infty} [F^{(j)}(t) - F^{(j+1)}(t)] \overline{D}(t) \, \mathrm{d}t, \tag{9.42}$$

and the expected number of backups before the full backup is

$$\sum_{j=1}^{\infty} j[G^{(j)}(K) - G^{(j+1)}(K)] \int_0^{\infty} \overline{D}(t) \, \mathrm{d}F^{(j+1)}(t)$$

$$+ \sum_{j=1}^{\infty} j G^{(j)}(K) \int_0^{\infty} [F^{(j)}(t) - F^{(j+1)}(t)] \, \mathrm{d}D(t)$$

$$= \sum_{j=1}^{\infty} G^{(j)}(K) \int_0^{\infty} \overline{D}(t) \, \mathrm{d}F^{(j)}(t). \tag{9.43}$$

Let us introduce the following costs: Cost c_F is incurred for the full backup, cost $c_K + c_0(x)$ is incurred for the cumulative backup when the total files are x ($0 \le x \le K$), and cost $c_R + c_0(x)$ is incurred for the recovery when the database fails, where $c_0(0) \equiv 0$. Using the same arguments for obtaining (9.26), the total expected cost until the full backup is

$$c_F + \sum_{j=0}^{\infty} \int_0^{\infty} [F^{(j)}(t) - F^{(j+1)}(t)] \, dD(t)$$

$$\times \left\{ \sum_{i=1}^{j} \int_0^K [c_K + c_0(x)] \, dG^{(i)}(x) + \int_0^K [c_R + c_0(x)] \, dG^{(j)}(x) \right\}$$

$$= c_F + \sum_{j=1}^{\infty} \int_0^{\infty} \overline{D}(t) \, dF^{(j)}(t) \int_0^K [c_K + c_0(x)] \, dG^{(j)}(x)$$

$$+ \sum_{j=0}^{\infty} \int_0^{\infty} [F^{(j)}(t) - F^{(j+1)}(t)] \, dD(t) \int_0^K [c_R + c_0(x)] \, dG^{(j)}(x). \quad (9.44)$$

Therefore, the expected cost rate is, dividing (9.44) by (9.42),

$$C_C(K) = \frac{\begin{array}{l} c_F + \sum_{j=1}^{\infty} \int_0^{\infty} \overline{D}(t) \, dF^{(j)}(t) \int_0^K [c_K + c_0(x)] \, dG^{(j)}(x) \\ + \sum_{j=0}^{\infty} \int_0^{\infty} [F^{(j)}(t) - F^{(j+1)}(t)] \, dD(t) \int_0^K [c_R + c_0(x)] \, dG^{(j)}(x) \end{array}}{\sum_{j=0}^{\infty} G^{(j)}(K) \int_0^{\infty} [F^{(j)}(t) - F^{(j+1)}(t)] \overline{D}(t) \, dt}.$$
$$(9.45)$$

In particular, when $K = 0$, i.e., the full backup is done at the first update or at the failure of the database, whichever occurs first, the expected cost in (9.45) is

$$C_C(0) = \frac{c_F + c_R \int_0^{\infty} \overline{F}(t) \, dD(t)}{\int_0^{\infty} \overline{F}(t) \overline{D}(t) \, dt}, \quad (9.46)$$

where $\overline{F}(t) \equiv 1 - F^{(1)}(t)$. When $K = \infty$, i.e., the full backup is done only at the failure of the database, the expected cost in (9.45) is

$$\frac{C_C(\infty)}{\gamma} = c_F + c_R + c_K \sum_{j=1}^{\infty} \int_0^{\infty} \overline{D}(t) \, dM_F(t)$$

$$+ \sum_{j=1}^{\infty} \int_0^{\infty} [2F^{(j)}(t) - F^{(j+1)}(t)] \, dD(t) \int_0^{\infty} c_0(x) \, dG^{(j)}(x), \quad (9.47)$$

where $M_F(t) \equiv \sum_{j=1}^{\infty} F^{(j)}(t)$.

Next, suppose that $c_0(x) = c_0 x$ and a database is updated in a Poisson process with rate $\alpha\lambda$, i.e., $F^{(j)}(t) - F^{(j+1)}(t) = [(\alpha\lambda t)^j / j!] e^{-\alpha\lambda t}$ ($j = 0, 1, 2, \cdots$), $D(t) = 1 - e^{-\beta\lambda t}$, and $\gamma = \beta\lambda$, where $0 < \alpha < 1$ and $\beta = 1 - \alpha$. In this case, the expected cost rate in (9.45) is rewritten as

$$\frac{C_C(K)}{\lambda} = \frac{c_F - c_K + (1 + \beta)c_0 \sum_{j=1}^{\infty} \alpha^j \int_0^K x \, dG^{(j)}(x)}{\sum_{j=0}^{\infty} \alpha^j G^{(j)}(K)} + c_K + \beta c_R. \quad (9.48)$$

We find an optimum level K^* that minimizes $C_C(K)$. Differentiating $C_C(K)$ with respect to K and setting it equal to zero,

$$\sum_{j=0}^{\infty} \alpha^j \int_0^K G^{(j)}(x)\,\mathrm{d}x = \frac{c_F - c_K}{(1+\beta)c_0}, \tag{9.49}$$

whose left-hand side is strictly increasing from 0 to ∞. Therefore, there exists an optimum K^* $(0 < K^* < \infty)$ that satisfies (9.49), and the resulting cost rate is

$$\frac{C_C(K^*)}{\lambda} = (1+\beta)c_0 K^* + c_K + \beta c_R. \tag{9.50}$$

Example 9.3. Suppose that $G(x) = 1 - \mathrm{e}^{-\mu x}$, i.e., $G^{(j)}(x) = \sum_{i=j}^{\infty}[(\mu x)^i/i!]\mathrm{e}^{-\mu x}$ $(j = 0, 1, 2, \cdots)$. Then, an optimum K^* is given by a unique solution of the equation

$$K - \frac{\alpha}{\beta\mu}(1 - \mathrm{e}^{-\beta\mu K}) = \frac{\beta}{1+\beta}\frac{c_F - c_K}{c_0}. \tag{9.51}$$

Furthermore, an optimum K^* is approximately

$$\widetilde{K} = \frac{1}{1+\beta}\frac{c_F - c_K}{c_0}, \tag{9.52}$$

and $K^* < \widetilde{K}$ that approaches $\widetilde{K}$, as $\beta \to 0$. In the same values of Example 9.2, $\mu K^* = 6.09$, $\mu\widetilde{K} = 23.53$, and $C_C(K^*)/(\lambda c_0/\mu) = 48.21$ ∎

Furthermore, when the full backup is done at time T before the total files exceed K or the database fails, and its full backup cost is c_F, the expected cost rate in (9.45) is easily extended as

$$C_C(K,T) = \frac{\begin{aligned}&c_F + \sum_{j=1}^{\infty}\int_0^T \overline{D}(t)\,\mathrm{d}F^{(j)}(t)\int_0^K [c_K + c_0(x)]\,\mathrm{d}G^{(j)}(x)\\ &+ \sum_{j=0}^{\infty}\int_0^T [F^{(j)}(t) - F^{(j+1)}(t)]\,\mathrm{d}D(t)\int_0^K [c_R + c_0(x)]\,\mathrm{d}G^{(j)}(x)\end{aligned}}{\sum_{j=0}^{\infty} G^{(j)}(K)\int_0^T [F^{(j)}(t) - F^{(j+1)}(t)]\overline{D}(t)\,\mathrm{d}t}. \tag{9.53}$$

When $c_0(x) = c_0 x$ and $K = \infty$, this corresponds to the cumulative backup model in Section 9.2.

Other Related Stochastic Models

The cumulative damage model is called the compound renewal process or the compound Poisson process in the theory of stochastic processes when shocks occur in a Poisson process. Examples to these processes of other practical fields are *total claims on an insurance company, drifting of stones on river beds, model for Brownian motion, distribution of galaxies, number of customers* or *amount of materials in a queuing process* or *storage process* [11, 238, 239] and *cancer epidemiology* [240, 241]. For example, we can apply the damage model to the simplest queuing process. A customer arrives at a counter with one server. If the server is free, the customer can be served immediately. Otherwise, if the server is busy with another customer, the customer has to wait for the service and forms a queue [61]. If the *arrivals of customers* are replaced with *shocks* and their *total times of waiting and service* with *total damage*, this corresponds to the cumulative damage model whose total damage decreases with time (Figure 10.1). In this process, we are mainly interested in the busy period that the server is working for arrival customers.

We introduce briefly typical related models such as the downtime of repairable systems, shot noise, insurance, and stochastic duels.

10.1 Other Models

(1) Downtime Distribution

An operating unit is repaired when it fails, and after the completion of its repair, it begins to operate again. It is assumed that the failure time is a random variable X_j having an identical distribution $F(t)$ with finite mean $1/\lambda$ and the repair time is a random variable variable W_j having an identical distribution $G(x)$ with finite mean $1/\mu$, *i.e.*, $F(t) \equiv \Pr\{X_j \le t\}$ and $G(x) \equiv \Pr\{W_j \le x\}$ $(j = 1, 2, \cdots)$. Then, the total downtime $D(t)$ during the interval $[0, t]$ is, replacing t in (2.3) with $t - x$ (see **(2)** of Section 2.1.1 in [1]),

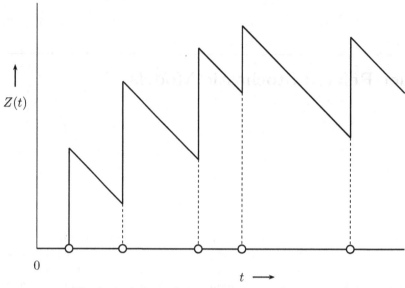

O Arrival time of customers

Fig. 10.1. Process for the total waiting and service time $Z(t)$ of a queuing model

$$\Pr\{D(t) \leq x\} = \sum_{j=0}^{\infty} G^{(j)}(x)[F^{(j)}(t-x) - F^{(j+1)}(t-x)], \qquad (10.1)$$

where $G^{(j)}(x)$ $(F^{(j)}(t))$ is the j-fold Stieltjes convolution of $G(x)$ $(F(t))$ with itself. Thus, the distribution that the total downtime exceeds a specified level $K > 0$ in time t is

$$\Pr\{D(t) > K\} = \sum_{j=0}^{\infty} [G^{(j)}(K) - G^{(j+1)}(K)]F^{(j)}(t-K) \quad \text{for } t > K.$$

The mean time that the total downtime first exceeds K is

$$\int_0^\infty \Pr\{D(t) \leq K\}\, dt = K + \frac{1}{\lambda}\left[\sum_{j=0}^{\infty} G^{(j)}(K)\right]. \qquad (10.2)$$

In particular, when $F(t) = 1 - e^{-\lambda t}$ and $G(x) = 1 - e^{-\mu x}$, from Example 2.2,

$\Pr\{D(t) > K\}$

$$= 1 - e^{-\lambda(t-K)}\left[1 + \sqrt{\lambda\mu(t-K)} \int_0^K e^{-\mu u} u^{-1/2} I_1(2\sqrt{\lambda\mu(t-K)u})\, du\right]$$

$$\text{for } t > K,$$

$$\int_0^\infty \Pr\{D(t) \le K\}\,dt = K + \frac{1}{\lambda}(1 + \mu K).$$

Next, let Y be the first time that one amount of downtime due to unit failures exceeds a fixed time $c > 0$, that is called an *allowed time*. Then, the distribution of a random variable Y and its mean time is, from (1.39) and (1.40) of [1], respectively,

$$\int_0^\infty e^{-st}\,d\Pr\{Y \le t\} = \frac{F^*(s)e^{-sc}\overline{G}(c)}{1 - F^*(s)\int_0^c e^{-st}\,dG(t)}, \tag{10.3}$$

$$E\{Y\} = \frac{1/\lambda + \int_0^c \overline{G}(t)\,dt}{\overline{G}(c)}, \tag{10.4}$$

where $\overline{G}(x) \equiv 1 - G(x)$, and $F^*(s)$ is the Laplace–Stieltjes (LS) transform of $F(t)$. The mean time $E\{Y\}$ is easily given by solving the renewal equation

$$E\{Y\} = \int_c^\infty \left(\frac{1}{\lambda} + c\right)\,dG(x) + \int_0^c \left(\frac{1}{\lambda} + x + E\{Y\}\right)\,dG(x).$$

(2) Shot Noise

Suppose that a shot noise occurs at time interval X_j and its amount is W_j. The total amount of shot noise is additive and falls into decay with time according to the rate function $h(\cdot)$. Then, the total amount of shot noise at time t is

$$Z(t) \equiv \sum_{j=1}^{N(t)} W_j h(t - S_j), \tag{10.5}$$

where $S_j \equiv \sum_{i=1}^j X_i$ and $N(t) \equiv \max_j\{S_j \le t\}$ [242, 243]. The stochastic behaviors of such shot noise were mathematically analyzed [244–248]. This can be also applied to riverflow [249], dams [250–253], and storage models [254–256]. If $h(t) = e^{-\alpha t}$, then this corresponds to the cumulative damage model with annealing in **(3)** of Section 2.5. Some failure distributions of reliability models were investigated by using the model of shot noise [126, 257].

(3) Insurance

The cumulative process can be applied to insurance, replacing *shock* with *claim* and *damage* with *claim size* [258]. In this case, random variables W_j, $N(t)$, and $Z(t)$ defined in (2.1) represent a claim size, the number of claims up to time t, and the total claim amount up to time t, respectively. Furthermore, the risk reserve $R(t)$ at time t is given by [259] (Figure 10.2)

$$R(t) = u + bt - \sum_{j=1}^{N(t)} W_j = u + bt - Z(t), \tag{10.6}$$

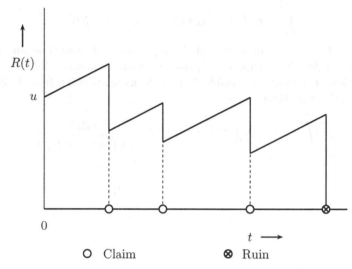

Fig. 10.2. Process for risk reserve $R(t)$ of an insurance model

where u is the initial risk reserve and $b > 0$ is the premium rate. The probability of ultimate ruin is given by

$$\psi(u) \equiv \Pr\{R(t) < 0 \text{ for some } t > 0\}$$
$$= \Pr\{Z(t) - bt > u \text{ for some } t > 0\}. \quad (10.7)$$

The properties of ruin probability $\psi(u)$ have been studied and summarized [258–261].

10.2 Stochastic Duels

This section introduces a classical model of stochastic duels in which each firing delivers an amount of damage governed by a random variable and it requires a specified threshold level of damage to kill the opponent. The theory of stochastic duels was studied [74, 75, 262–266]. The optimum engagement problem of shooting strategy with incomplete damage information was considered [267].

The stochastic model in which each firing delivers the same amount of damage to the opponent and the kill requires a fixed number of hits was proposed, and the probability that a duelist wins against the opponent was obtained [263, 264]. In addition, the weapon lifetimes that can be functions of time or number of rounds fired were considered [265], and the total damage resulting from firings was assumed to depend on both time and the number of rounds fired [75]. Recently, multiple damage functions to estimate the probability that a single weapon detonation destroys a point target were discussed [266].

This section assumes that each firing delivers an amount of damage and it requires a prespecified threshold level of damage to the opponent, where each damage is additive. A duelist loses when the total damage exceeds a threshold level. This corresponds to the cumulative damage model by replacing *rounds fired* with *shocks* and *threshold level* with *failure level*.

We consider five models of stochastic duels and derive analytically the probabilities of winning the duel with reference to Chapter 2.

(1) Standard Model

Consider a stochastic duel with two contestants, say, A and B. Both contestants have unlimited ammunition and unlimited time to kill the opponent. Duelist A (B) begins simultaneously with a weapon and fires at time intervals according to an identical probability distribution $F_A(t)$ with finite mean $1/\lambda_A$ ($F_B(t)$ with finite mean $1/\lambda_B$), respectively, *i.e.*, $F_A(t)$ and $F_B(t)$ are distribution functions of times between rounds fired. Each firing delivers an amount of damage with a general distribution $G_A(x)$ ($G_B(x)$), and requires a threshold level K_A (K_B) of the total damage to kill the opponent. Duelist A (B) wins the duel if he or she delivers K_A (K_B) to A (B), respectively. It is assumed that each damage is additive and does not deteriorate.

Let $Z_A(t)$ ($Z_B(t)$) be the total damage up to time t by A (B). Recalling that duelist A kills B when the total damage delivered by A exceeds a threshold level K_A, the probability that A kills B up to time t is, from (2.9),

$$\Phi_A(t) \equiv \Pr\{Z_A(t) > K_A\} = \sum_{j=0}^{\infty} [G_A^{(j)}(K_A) - G_A^{(j+1)}(K_A)] F_A^{(j+1)}(t). \quad (10.8)$$

Taking the LS transform of (10.8),

$$\Phi_A^*(s) \equiv \int_0^{\infty} e^{-st}\, d\Phi_A(t) = \sum_{j=0}^{\infty} [G_A^{(j)}(K_A) - G_A^{(j+1)}(K_A)][F_A^*(s)]^{j+1}, \quad (10.9)$$

where $F_A^*(s)$ is the LS transform of $F_A(t)$. The mean time for A to kill B is

$$l_A \equiv \int_0^{\infty} t\, d\Phi_A(t) = \frac{1}{\lambda_A} \sum_{j=0}^{\infty} G_A^{(j)}(K_A). \quad (10.10)$$

In the same fashion, the probability $\Phi_B(t)$ that B kills A up to time t can be obtained by exchanging from suffix A into B.

Therefore, the probability $P_A(t)$ that A wins the duel up to time t is

$$P_A(t) = \int_0^t [1 - \Phi_B(u)]\, d\Phi_A(u), \quad (10.11)$$

and conversely, the probability $P_B(t)$ that B wins the duel up to time t is

$$P_B(t) = \int_0^t [1 - \Phi_A(u)] \, d\Phi_B(u). \tag{10.12}$$

(2) Imperfect Hit

It is assumed that A (B) hits the opponent B (A) with probability p_A (p_B) and A (B) misses B (A) with $q_A \equiv 1 - p_A$ ($q_B \equiv 1 - p_B$), respectively. Then, the probability distribution of time for A to score one hit on B up to time t is, from Example 1.1,

$$F_1(t) = [1 + q_A F_A(t) + q_A F_A(t) * q_A F_A(t) + \cdots] * p_A F_A(t).$$

Thus, replacing $F_A(t)$ in (10.8) with $F_1(t)$, we have $\Phi_A(t)$. The LS transform is

$$\Phi_A^*(s) = \sum_{j=0}^{\infty} [G_A^{(j)}(K_A) - G_A^{(j+1)}(K_A)] \left[\frac{p_A F_A^*(s)}{1 - q_A F_A^*(s)} \right]^{j+1}, \tag{10.13}$$

and the mean time for A to kill B is

$$l_A = \frac{1}{p_A \lambda_A} \sum_{j=0}^{\infty} G_A^{(j)}(K_A). \tag{10.14}$$

The other quantities can be obtained in a similar fashion.

(3) Independent Damage

It is assumed that the amount of damage is not additive and the amount is nullified immediately when it is less than K_A (K_B). The other assumptions are the same as those of case (1) except that the total damage is additive. Then, the LS transform of the probability that A kills B up to time t is, from Section 2.2,

$$\Phi_A^*(s) = \sum_{j=0}^{\infty} \left\{ [G_A(K_A)]^j - [G_A(K_A)]^{j+1} \right\} [F_A^*(s)]^{j+1}$$

$$= \frac{[1 - G_A(K_A)] F_A^*(s)}{1 - G_A(K_A) F_A^*(s)}, \tag{10.15}$$

and the mean time for A to kill B is

$$l_A = \frac{1}{\lambda_A [1 - G_A(K_A)]}. \tag{10.16}$$

(4) Random Threshold Level

It is assumed that a threshold level K_A (K_B) is a random variable with a general distribution $L_A(x)$ $(L_B(x))$, respectively. Then, from **(2)** in Section 2.5, for case **(1)**,

$$\Phi_A^*(s) = \sum_{j=0}^{\infty} [F_A^*(s)]^{j+1} \int_0^{\infty} [G_A^{(j)}(x) - G_A^{(j+1)}(x)] \, dL_A(x), \qquad (10.17)$$

for case **(2)**,

$$\Phi_A^*(s) = \sum_{j=0}^{\infty} \left[\frac{p_A F_A^*(s)}{1 - q_A F_A^*(s)} \right]^{j+1} \int_0^{\infty} [G_A^{(j)}(x) - G_A^{(j+1)}(x)] \, dL_A(x), \quad (10.18)$$

and for case **(3)**,

$$\Phi_A^*(s) = \sum_{j=0}^{\infty} [F_A^*(s)]^{j+1} \int_0^{\infty} \left\{ [G_A(x)]^j - [G_A(x)]^{j+1} \right\} \, dL_A(x). \qquad (10.19)$$

The other quantities can be obtained in a similar fashion.

(5) Lifetimes of Weapons

Consider the lifetimes of A's (B's) weapon distributed with $R_A(t)$ $(R_B(t))$, respectively. It is assumed that the failed weapon of A (B) remains in the duel until A (B) is killed or B's (A's) weapon fails. Then, the probability that A wins in the duel up to time t is

$$P_A(t) = \int_0^t [1 - R_A(u)] \left\{ 1 - \int_0^u [1 - R_B(v)] \, d\Phi_B(v) \right\} d\Phi_A(u), \qquad (10.20)$$

and the tie probability is

$$P_{AB}(t) = \int_0^t [1 - \Phi_A(u)] \, dR_A(u) \int_0^t [1 - \Phi_B(u)] \, dR_B(u), \qquad (10.21)$$

that represents the probability that both A and B cannot kill the opponent because of failures of the weapons up to time t. Note that $P_A(\infty) + P_B(\infty) + P_{AB}(\infty) = 1$.

Example 10.1. It is assumed that $G_A(x) \equiv 0$ for $x < 1$ and 1 for $x \geq 1$ and K_A is a positive integer. Then, from (10.13),

$$\Phi_A^*(s) = \left[\frac{p_A F_A^*(s)}{1 - q_A F_A^*(s)} \right]^{K_A}.$$

Furthermore, $L_A(x)$ is a discrete distribution, *i.e.*,

$$\Pr\{K_A = j\} = \alpha_j \quad (j = 1, 2, \cdots),$$

where $\sum_{j=1}^{\infty} \alpha_j = 1$. Then, from (10.18),

$$\Phi_A^*(s) = \sum_{j=1}^{\infty} \alpha_j \left[\frac{p_A F_A^*(s)}{1 - q_A F_A^*(s)} \right]^j. \quad \blacksquare$$

Example 10.2. Suppose in case (4) that all random variables are exponential, i.e., $F(t) = 1 - e^{-\lambda t}$, $G(x) = 1 - e^{-\mu x}$, and $L(x) = 1 - e^{-\alpha x}$, where the suffixes of the three parameters are omitted. Then, from (10.18),

$$\Phi_A^*(s) = \frac{\alpha \lambda p}{(\alpha + \mu)s + \alpha \lambda p}.$$

By inversion,

$$\Phi_A(t) = 1 - e^{-\theta_A t},$$

where $\theta_A \equiv \alpha \lambda p / (\alpha + \mu)$. For duelist B,

$$\Phi_B(t) = 1 - e^{-\theta_B t}.$$

Thus, from (10.11),

$$P_A(t) = \frac{\theta_A}{\theta_A + \theta_B} [1 - e^{-(\theta_A + \theta_B)t}],$$

$$P_B(t) = \frac{\theta_B}{\theta_A + \theta_B} [1 - e^{-(\theta_A + \theta_B)t}].$$

Furthermore, when the lifetimes of the weapons are assumed to be $R_A(t) = 1 - e^{-\gamma_A t}$ and $R_B(t) = 1 - e^{-\gamma_B(t)}$, from case (5),

$$P_A(t) = \frac{\theta_A}{\gamma_B + \theta_B} \left\{ \frac{\gamma_B}{\gamma_A + \theta_A} [1 - e^{-(\gamma_A + \theta_A)t}] \right.$$
$$\left. + \frac{\theta_B}{\gamma_A + \theta_A + \gamma_B + \theta_B} [1 - e^{-(\gamma_A + \theta_A + \gamma_B + \theta_B)t}] \right\},$$

$$P_B(t) = \frac{\theta_B}{\gamma_A + \theta_A} \left\{ \frac{\gamma_A}{\gamma_B + \theta_B} [1 - e^{-(\gamma_B + \theta_B)t}] \right.$$
$$\left. + \frac{\theta_A}{\gamma_A + \theta_A + \gamma_B + \theta_B} [1 - e^{-(\gamma_A + \theta_A + \gamma_B + \theta_B)t}] \right\},$$

$$P_{AB}(t) = \frac{\gamma_A}{\gamma_A + \theta_A} \frac{\gamma_B}{\gamma_B + \theta_B} [1 - e^{-(\gamma_A + \theta_A)t}][1 - e^{-(\gamma_B + \theta_B)t}],$$

where it is clearly seen that $P_A(\infty) + P_B(\infty) + P_{AB}(\infty) = 1$. $\blacksquare$

References

1. Nakagawa T (2005) Maintenance Theory of Reliability. Springer, London.
2. Bogdanoff JL, Kozin F (1985) Probabilistic Models of Cumulative Damage. John Wiley & Sons, New York.
3. Hudson WR, Haas R, Uddin W (1997) Infrastructure Management. McGraw-Hill, New York.
4. Hisano K (2000) Preventive maintenance and residual life evaluation technique for power plant–Preventive maintenance. Thermal Nucl Power 51:491–517.
5. Hisano K (2001) Preventive maintenance and residual life evaluation technique for power plant–Review of future advances in preventive maintenance technology. Thermal Nucl Power 52:363–370.
6. Durham SD, Padgett WJ(1990) Estimation for a probabilistic stress-strength model. IEEE Trans Reliab 39:199–203.
7. Miner MA (1945) Cumulative damage in fatigue. J Appl Mech 12:A159–A164.
8. Birnbaum Z, Saunders SC (1968) A probabilistic interpretation of Miner's rule. SIAM J Appl Math 16:637–652.
9. Stallmeyer JE, Walker WH (1968) Cumulative damage theories and application. J Struct Div ASCE 94:2739–2750.
10. Kuroishi T, Minami Y, Kobayashi Y, Yokoyama T, Hasegawa Y, Kageyama O, Minatomoto M (2003) Power systems:A portal to customer services for electric power generation. Mitsubishi Heavy Industries, Ltd. Tech Rev 40:1–9.
11. Cox DR (1962) Renewal Theory. Methuen, London.
12. Akama M, Ishizuka H (1995) Reliability analysis of Shinkansen vehicle axle using probabilistic fracture mechanics. JSME Inter J Ser A 38:378–383.
13. Gertsbakh I, Kordonsky Kh (1969) Models of Failure. Springer, Berlin.
14. Satow T, Teramoto K, Nakagawa T (2000) Optimal replacement policy for a cumulative damage model with time deterioration. Math Comput Model 31:313–319.
15. Durham SD, Padgett WJ (1997) Cumulative damage models for system failure with application to carbon fibers and composites. Technometrics 39:34–44.
16. Padgett WJ (1998) A muliplicative damage model for strength of fibrous composite materials. IEEE Trans Reliab 47:46–52.
17. Ihara C, Tsurui A (1977) Fatigue of metals as stochastic phenomena. J Eng Mater Tech 99:26–28.

18. Sobczyk K, Trebicki J (1989) Modelling of random fatigue by cumulative jump processes. Eng Fracture Mech 34:477–493.
19. Scarf PA, Wang W, Laycock PJ (1996) A stochastic model of crack growth under periodic inspections. Reliab Eng Syst Saf 51:331–339.
20. Hopp WJ, Kuo YL (1998) An optimal structured policy for maintenance of partially observable aircraft engine components. Nav Res Logist 45:335–352.
21. Lukić M, Cremona C (2001) Probabilistic optimization of welded joints maintenance versus fatigue and fracture. Reliab Eng Syst Saf 72:253–264.
22. Garbatov Y, Soares CG (2001) Cost and reliability based strategies for fatigue maintenance planning of floating structures. Reliab Eng Syst Saf 73:293–301.
23. Petryna YS, Pfanner D, Stangenberg F, Krätzig WB (2002) Reliability of reinforced concrete structures under fatigue. Reliab Eng Syst Saf 77:253–261.
24. Campean IF, Rosala GF, Grove DM, Henshall E (2005) Life modelling of a plastic automotive component. In:Proc Ann Reliab Maintainability Symp:319–325.
25. Sobczyk K (1987) Stochastic models for fatigue damage of materials. Adv Appl Probab 19:652–673.
26. Sobczyk K and Spencer Jr BF (1992) Random Fatigue:From Data to Theory. Academic, New York.
27. Dasgupta A, Pecht M (1991) Material failure mechanisms and damage models. IEEE Trans Reliab 40:531–536.
28. Smith WL (1955) Regenerative stochastic processes. Proc Roy Soc London A 232:6–31.
29. Smith WL (1958) Renewal theory and its ramifications. J Roy Statist Soc B 20:243–302.
30. Mercer A (1961) On wear-dependent renewal processes. J Roy Statist Soc B 23:368–376.
31. Morey RC (1966) Some stochastic properties of a compound-renewal damage model. Oper Res 14:902–908.
32. Murthy VK, Lients BP (1968) On cumulative damage and reliability of components. Document ARL68–0180, Aerospace Res Lab, Wright-Patterson Air Force Base, Ohio.
33. Esary JD, Marshall AW, Proschan F (1973) Shock models and wear processes. Ann Probab 1:627–649.
34. Gaver Jr DP (1963) Random hazard in reliability problems. Technometrics 5:211–226.
35. Antelman G, Savage IR (1965) Characteristic functions of stochastic integrals and reliability theory. Nav Res Logist Q 12:199–222.
36. Birnbaum ZW, Saunders SC (1969) A new family of life distributions. J Appl Probab 6:319–327.
37. Reynolds DS, Savage IR (1971) Random wear models in reliability theory. Adv Appl Probab 3:229–248.
38. Colombo AG, Reina G, Volta G (1974) Extreme value characteristics of distributions of cumulative processes. IEEE Trans Reliab R-23:179–186.
39. Aven T, Jensen U (1999) Stochastic Models in Reliability. Springer, New York.
40. Nakagawa T, Osaki S (1974) Some aspects of damage model. Microelectron Reliab 13:253–257.
41. Rausand M, Høyland A (2004) System Reliability Theory. John Wiley & Sons, Hoboken NJ.

42. Taylor HM (1975) Optimal replacement under additive damage and other failure models. Nav Res Logist Q 22:1–18.
43. Feldman RM (1976) Optimal replacement with semi-Markov shock models. J Appl Probab 13:108–117.
44. Feldman RM (1977) Optimal replacement with semi-Markov shock models using discounted costs. Math Oper Res 2:78–90.
45. Feldman RM (1977) Optimal replacement for systems governed by Markov additive shock processes. Ann Probab 5:413–429.
46. Zuckerman D (1977) Replacement models under additive damage. Nav Res Logist Q 24:549–558.
47. Zuckerman D (1978) Optimal replacement policy for the case where the damage process is a one–sided Lévy process. Stoch Process Appl 7:141–151.
48. Zuckerman D (1978) Optimal stopping in a semi-Markov shock model. J Appl Probab 15:629–634.
49. Zuckerman D (1980) Optimal replacement under additive damage and self-restoration. RAIRO Oper Res 14:115–127.
50. Zuckerman D (1980) A note on the optimal replacement time of damaged devices. Nav Res Logist Q 27:521–524.
51. Nakagawa T (1976) On a replacement problem of a cumulative damage model. Oper Res Q 27:895–900.
52. Nakagawa T (1979) Replacement problem of a parallel system in random environment. J Appl Probab 16:203–205.
53. Nakagawa T (1979) Further results of replacement problem of a parallel system in random environment. J Appl Probab 16:923–926.
54. Nakagawa T, Murthy DNP (1993) Optimal replacement policies for a two-unit system with failure interactions. RAIRO Oper Res 27:427–438.
55. Nakagawa T, Kijima M (1989) Replacement policies for a cumulative damage model with minimal repair at failure. IEEE Trans Reliab 38:581–584.
56. Nakagawa T (1986) Modified discrete preventive maintenance policies. Nav Res Logist Q 33:703–715.
57. Kijima M, Nakagawa T (1992) Replacement policies of a shock model with imperfect preventive maintenance. Eur J Oper Res 57:100–110.
58. Satow T, Yasui K, Nakagawa T (1996) Optimal garbage collection policies for a database in a computer system:RAIRO Oper Res 30:359–372.
59. Qian CH, Nakamura S, Nakagawa T, (1999) Cumulative damage model with two kinds of shocks and its application to the backup policy. J Oper Res Soc Jpn 42:501–511.
60. Ross SM (1983) Stochastic Processes. John Wiley & Sons, New York.
61. Osaki S (1992) Applied Stochastic System Modeling. Springer, Berlin.
62. Karlin S, Taylor HM (1975) A First Course in Stochastic Processes. Academic, New York.
63. Çinlar E (1975) Introduction to Stochastic Processes. Prentice-Hall, Englewood Cliffs, NJ.
64. Trindade D, Nathan S (2005) Simple plots for monitoring the field reliability of repairable systems. In:Proc Ann Reliab Maintainability Symp:539–544.
65. Barlow RE, Proschan F (1965) Mathematical Theory of Reliability. John Wiley & Sons, New York.
66. Nakagawa T, Kowada M (1983) Analysis of a system with minimal repair and its application to replacement policy. Eur J Oper Res 12:176–182.

67. Barbour AD, Chryssaphinou O (2001) Compound Poisson approximation:A user's guide. Ann Appl Probab 11:964–1002.
68. Abdel-Hameed M, Proschan F (1975) Shock models with underlying birth process. J Appl Probab 12:18–28.
69. Klefsjö B (1981) Survival under the pure birth shock model. J Appl Probab 18:554–560.
70. Abdel-Hameed M (1984) Life distribution properties of devices subject to a Lévy wear process. Math Oper Res 9:606–614.
71. Khoshnevisan D, Xiao Y (2002) Level sets of additive Lévy processes. Ann Probab 30:62–100.
72. Block HW, Savits TH (1978) Shock models with NBUE survival. J Appl Probab 15:621–628.
73. Pellerey F (1994) Shock models with underlying counting process. J Appl Probab 31:156–166.
74. Ancker Jr CJ (1967) The status of developments in the theory of stochastic duels–II. Oper Res 15:388–406.
75. Nagabhushanam A, Jain GC (1972) Stochastic duels with damage. Oper Res 20:350–356.
76. Råde L (1976) Reliability systems in random environment. J Appl Probab 13:407–410.
77. Hokstad P (1988) A shock model for comnon-cause failures. Reliab Eng Syst Saf 23:127–145.
78. Vaurio JK (1994) The theory and quantification of common cause shock events for redundant standby systems. Reliab Eng Syst Saf 43:289–305.
79. Vaurio JK (1995) The probability modeling of external common cause failure shocks in redundant systems. Reliab Eng Syst Saf 50:97–107.
80. Kvam PH, Martz HF (1995) Bayesian inference in a discrete shock model using confounded common cause data. Reliab Eng Syst Saf 48:19–25.
81. Abdel-Hameed M (1984) Life distribution properties of devices subject to a pure jump damage process. J Appl Probab 21:816–825.
82. Grandell J (1976) Doubly Stochastic Poisson Process. Lecture Notes in Mathematics 529. Springer, New York.
83. Takács L (1960) Stochastic Processes. John Wiley & Sons, New York.
84. Gut A (1990) Cumulative shock models. Adv Appl Probab 22:504–507.
85. Glynn PW, Whitt W (1993) Limit theorems for cumulative processes. Stoch Process Appl 47:299–314.
86. Roginsky AL (1994) A central limit theorem for cumulative processes. Adv Appl Probab 26:104–121.
87. Finkelstein MS, Zarudnij VI (2001) A shock process with a non-cumulative damage. Reliab Eng Syst Saf 71:103–107.
88. Barlow RE, Proschan F (1975) Statistical Theory of Reliability and Life Testing. Holt, Rinehart & Winston, New York.
89. Abdel-Hameed M and Proschan F (1973) Nonstationary shock models. Stoch Process Appl 1:383–404.
90. Shaked M (1984) Wear and damage processes from shock models in reliability theory. In:Abdel-Hameed M, Çinlar E, Quinn J (eds) Reliability Theory and Models. Academic, Orlando.
91. Severina TI (1975) A model of accumulation of damage. Eng Cybern 13:74–76.
92. Ramanarayanan R (1976) Cumulative damage processes and alertness of the worker. IEEE Trans Reliab R-25:281–283.

93. Gottlieb G (1980) Failure distributions of shock models. J Appl Probab 17:745–752.

94. Thall PF (1981) Cluster shock models. J Appl Probab 18:104–111.

95. Klefsjö B (1981) HNBUE survival under some shock models. Scand J Stat 8:39–47.

96. Ross SM (1981) Generalized Poisson shock models. Ann Probab 9:896–898.

97. Neuts MF, Bhallacharjee MC (1981) Shock models with phase-type survival and shock resistance. Nav Res Logist Quart 28:213–219.

98. Ghosh M, Ebrahimi N (1982) Shock models leading to increasing failure rate and decreasing mean residual life survival. J Appl Probab 19:158–166.

99. Ohi F, Nishida T (1983) Another proof of IFRA property of S.M. Ross' generalized Poisson shock models. Math Jpn 28:117–123.

100. Ebrahimi N (1985) A stress-strength system. J Appl Probab 22:467–472.

101. Yamada K (1989) Limit theorems for jump shock models. J Appl Probab 27:793–806.

102. Singh H, Jain K (1989) Preservation of some partial ordering under Poisson shock models. Adv Appl Probab 21:713–716.

103. Kochar SC (1990) On preservation of some partial ordering under shock models. Adv Appl Probab 22:508–509.

104. Manoharan M, Singh H, Misra N (1992) Preservation of phase-type distributions under Poisson shock models. Adv Appl Probab 24:223–225.

105. Pellerey F (1993) Partial orderings under cumulative damage shock models. Adv Appl Probab 25:939–946.

106. Fagiuoli E, Pellerey F (1994) Preservation of certain classes of life distributions under Poisson shock models. J Appl Probab 31:458–465.

107. Ebrahimi N (1999) Stochastic properties of a cumulative damage threshold crossing model. J Appl Probab 36:720–732.

108. Shanthikumar JG, Sumita U (1983) General shock models associated with correlated renewal sequences. J Appl Probab 20:600–614.

109. Sumita U, Shanthikumar JG (1985) A class of correlated shock models. Adv Appl Probab 17:347–366.

110. Anderson KK (1987) Limit theorems for general shock models with infinite mean intershock times. J Appl Probab 24:449–456.

111. Anderson KK (1988) A note on cumulative shock models. J Appl Probab 25:220–223.

112. Pérez-Ocón R, Gámiz-Pérez ML (1995) On the HNBUE property in a class of correlated cumulative shock models. Adv Appl Probab 27:1186–1188.

113. Igaki N, Sumita U, Kowada M (1995) Analysis of Markov renewal shock models. J Appl Probab 32:821–831.

114. Li G, Luo J (2005) Shock model in Markovian environment. Nav Res Logist 52:253–260.

115. Marchall AW, Shaked M (1979) Mulitivariate shock models for distributions with increasing hazard rate average. Ann Probab 7:343–358.

116. Ohi F, Nishida T (1979) Bivariate shock models:NBU and NBUE properties, and positively quadrand dependency. J Oper Res Soc Jpn 22:266–273.

117. Savits TH, Shaked M (1981) Shock models and the MIFRA property. Stoch Process Appl 11:273–283.

118. Griffith WS (1982) Remarks on univariate shock model with some bivariate generation. Nav Res Logist Q 29:63–74.

119. Shaked M, Shanthikumar JG (1987) IFRA properties of some Markov jump processes with general state space. Math Oper Res 12:562–568.
120. Savits TH (1988) Some multivariate distributions derived from a non-fatal shock model. J Appl Probab 25:383–390.
121. Wong T (1997) Preservation of multivariate stochastic orders under multivariate Poisson shock models. J Appl Probab 34:1009–1020.
122. Mallor F, Omey E (2001) Shocks, runs and random sums. J Appl Probab 38:438–448.
123. Belzunce F, Lillo RE, Pellerey F, Shaked M (2002) Preservation of association in multivariate shock and claim models. Oper Res Lett 30:223–230.
124. Gaudoin O, Soler JL (1997) Failure rate behavior of components subjected to random stresses. Reliab Eng Syst Saf 58:19–30.
125. Abdel-Hameed M (1975) A gamma wear process. IEEE Trans Reliab R-24:152–153.
126. Lemoine AJ, Wenocur ML(1985) On failure modeling. Nav Res Logist Q 32:497–508.
127. Desmond A (1985) Stochastic models of failure in random environments. Can J Stat 13:171–183.
128. Kececioglu DB, Jiang MX (1998) A unified approach to random-fatigue reliability quantification under random loading. In:Proc Ann Reliab Maintainability Symp:308–313.
129. Owen WJ, Padgett WJ (2003) Accelerated test models with the Birnbaum-Saunders distribution. In:Pham H (ed) Handbook of Reliability Engineering. Springer, London:429–439.
130. Park C, Padgett WJ (2005) New cumulative damage models for failure using stochastic processes as initial damage. IEEE Trans Reliab 54:530–540.
131. Satow T, Yasui K, Nakagawa T (1996) Optimal garbage collection policies for a database with random threshold level. Electron Commun Japan 79:31–40.
132. Nakagawa T (1975) On cumulative damage with annealing. IEEE Trans Reliab R-24:90–91.
133. Nakagawa T (1976) On a cumulative damage model with N different components. IEEE Trans Reliab R-25:112–114.
134. Satow T, Nakagawa T (1997) Three replacement models with two kinds of damage. Microelectron Reliab 37:909–913.
135. Satow T, Nakagawa T (1997) Replacement policies for a shock model with two kinds of damage. In:Osaki S (ed) Stochastic Modelling in Innovative Manufacturing, Springer Lecture Notes in Economics and Mathematical Systems 445:188–195.
136. Qian CH, Ito K, Nakagawa T (2005) Optimal preventive maintenance policies for a shock model with given damage level. J Qual Maint Eng 11:216–227.
137. Gottlieb G, Yechiali U (1980) Damage models for multi-component systems. Eur J Oper Res 5:193–197.
138. Bergman B (1978) Optimal replacement under a general failure model. Adv Appl Probab 10:431–451.
139. Abdel-Hameed M, Shimi IN (1978) Optimal replacement of damaged devices. J Appl Probab 15:153–161.
140. Yamada K (1980) Explicit formula of optimal replacement under additive shock processes. Stoch Process Appl 9:193–208.
141. Chikte SD, Deshmukh SD (1981) Preventive maintenance and replacement under additive damage. Nav Res Logist Q 28:33–46.

142. Gottlieb G (1982) Optimal replacement for shock models with general failure rate. Oper Res 30:82–92.

143. Waldmann KH (1983) Optimal replacement under additive damage in randomly varying environments. Nav Res Logist Q 30:377–386.

144. Gottlieb G, Levikson B (1984) Optimal replacement for self-repairing shock models with general failure rate. J Appl Probab 21:108–119.

145. Mizuno N (1986) Generalized mathematical programming for optimal replacement in a semi-Markov shock model. Oper Res 34:790–795.

146. Aven T, Gaarder S (1987) Optimal replacement in a shock model:Discrete time. J Appl Probab 24:281–287.

147. Abdel-Hameed M, Nakhi Y (1991) Optimal replacement and maintenance of systems subject to semi-Markov damage. Stoch Process Appl 37:141–160.

148. Abdel-Hameed M (1999) Applications of semi-Markov processes in reliability and maintenance. In:Jenssen J, Limnios N (eds) Semi-Markov Models and Applications. Kluwer Academic, The Netherlands:337–348.

149. Posner MJM, Zuckerman D (1984) A replacement model for an additive damage model with restoration. Oper Res Lett 3:141–148.

150. Posner MJM, Zuckerman D (1986) Semi-Markov shock models with additive damage. Adv Appl Probab 18:772–790.

151. Perry D, Posner MJM (1991) Determining the control limit policy in a replacement model with restoration. Oper Res Lett 10:335–341.

152. Perry D (2000) Control limit policies in a replacement model with additive phase-type distributed damage and linear restoration. Oper Res Lett 27:127–134.

153. Wortman MA, Klutke GA, Ayhan H (1994) A maintenance strategy for systems subjected to deterioration governed by random shocks. IEEE Trans Reliab 43:439–445.

154. Feng W, Adachi K, Kowada M (1994) Dynamically optimal replacement policy for a shock model in a Markov random environment. J Oper Res Soc Jpn 37:255–270.

155. Aven T (1996) Condition based replacement policies–A counting process approach. Reliab Eng Sys Saf 51:275–281.

156. Sheu SH, Griffith WS (1996) Optimal number of minimal repairs before replacement of a system subject to shocks. Nav Res Logist 43:319–333.

157. Sheu SH (1997) Extend block replacement policy of a system subject to shocks. IEEE Trans Reliab 46:375–382.

158. Sheu SH (1998) A generalized age and block replacement of a system subject to shocks. Eur J Oper Res 108:345–362.

159. Sheu SH, Griffith WS (2002) Extend block replacement policy with shock models and used items. Eur J Oper Res 140:50–60.

160. Sheu SH, Chien YH (2004) Optimal age-replacement policy of a system subject to shocks with random lead-time. Eur J Oper Res 159:132–144.

161. Wang GJ, Zhang YL (2005) A shock model with two-type failures and optimal replacement policy. Int J Syst Sci 36:209–214.

162. Feldman RM, Joo NY (1985) A state-age dependent policy for a shock process. Stoch Models 1:53–76.

163. Lam CT, Yeh RH (1994) Optimal replacement policies for multi-state deteriorating systems. Nav Res Logist 41:303–315.

164. Klutke GA, Yang YJ (2002) The availability of inspected systems subject to shocks and graceful degradation. IEEE Trans Reliab 51:371–374.

165. Li WJ, Pham H (2005) Reliability modeling of multi-state degraded systems with multi-competing failures and random shocks. IEEE Trans Reliab 54:297–303.

166. Li Z, Chan LY, Yuan Z (1999) Failure time distribution under a δ-shock model and its application economic design of systems. Inter J Reliab Qual Saf Eng 6:237–247.

167. Tang YY, Lam Y (2006) A δ-shock maintenance model for a deteriorating system. Eur J Oper Res 168:541–556.

168. Qian CH, Nakamura S, Nakagawa T (2000) Replacement policies for cumulative damage model with maintenance cost. Scientiae Mathematicae 3:117–126.

169. Muth E (1977) An optimal decision rule for repair vs. replacement. IEEE Trans Reliab 26:179–181.

170. Barbera F, Schneider H, Watson E (1999) A condition based maintenance model for a two-unit series system. Eur J Oper Res 116:281–290.

171. Murthy DNP, Nguyen DG (1985) Study of two-component system with failure interaction. Nav Res Logist Q 32:239–248.

172. Pham H, Suprasad A, Misra B (1996) Reliability and MTTF prediction of k-out-of-n complex systems with components subjected to multiple stages of degradation. Int J Syst Sci 27:995–1000.

173. Skoulakis G (2000) A general shock model for a reliability system. J Appl Probab 37:925–935.

174. Juang MG, Sheu SH (2003) Graphical approach to replacement policy of a K-out-of-N system subject to shocks. Int J Reliab Qual Saf Eng 10:55–68.

175. Chryssaphinou O, Papastavridis S (1990) Reliability of a consecuitive-k-out-of-n system in a random environment. J Appl Probab 27:452–458.

176. Petakos K, Tsapelas T (1997) Reliability analysis for systems in a random environment. J Appl Probab 34:1021–1031.

177. Shaked M, Shantikumar JG (1990) Reliability and maintainability. In:Heyman DP, Sobel MJ(eds) Stochastic Models. North Holland, Amsterdam.

178. Murthy DNP, Casey RT (1987) Optimal policy for a two component system with shock type failure interaction. In:Proc 8th Nat Conf Ausl Oper Res Soc, Melbourne, 161–172.

179. Murthy DNP, Nguyen DG (1985) Study of a multi-component system with failure interaction. Eur J Oper Res 21:330–338.

180. Murthy DNP, Wilson RJ (1994) Parameter estimation in multi-component with failure interaction. Stoch Models Data Analysis 10:47–60.

181. Jhang JP, Sheu SH (2000) Optimal age and block replacement policies for a multi-component system with failure interaction. Inter J Syst Sci 31:593–603.

182. Scarf PA, Deara M (2003) Block replacement policies for a two-component system with failure dependence. Nav Res Logist 50:70–87.

183. Zequeria RI, Bérenguer C (2004) Maintenance cost analysis of a two-component parallel system with failure interaction. In:Proc Ann Reliab Maintainability Symp:220-225.

184. Zequeira RI, Bérenguer C (2005) On the inspection policy of a two-component parallel system with failure interaction. Reliab Eng Syst Saf 88:99–107.

185. Satow T, Osaki S (2003) Optimal replacement policies for a two-unit system with shock damage interaction. Comput Math Appl 46:1129–1138.

186. Boland PJ, Proschan F (1983) Optimal replacement of a system subject to shocks. Oper Res 31:697–704.

187. Abdel-Hameed M (1986) Optimum replacement of a system subject to shocks. J Appl Probab 23:107–114.

188. Puri PS, Singh H (1986) Optimum replacement of a system subject to shocks:A mathematical lemma. Oper Res 34:782–789.

189. Rangan A, Grace RE (1988) A non-Markov model for the optimum replacement of self-replacement systems subject to shocks. J Appl Probab 25:375–382.

190. Nakagawa T (1987) Modified, discrete replacement models. IEEE Trans Reliab 36:243–245.

191. Satow T, Nakagawa T (1997) Optimal replacement policy for a cumulative damage model with deteriorated inspection. Int J Reliab Qual Saf Eng 4:387–393.

192. Park KS (1988) Optimal continuous-wear limit replacement under periodic inspections. IEEE Trans Reliab 37:97–102.

193. Park KS (1988) Optimal wear-limit replacement with wear-dependent failures. IEEE Trans Reliab 37:293–294.

194. Grall A, Bérenguer C, Dieulle L (2002) A condition-based maintenance policy for stochastically deterioration systems. Reliab Eng Syst Saf 76:167–180.

195. Dieulle L, Bérenguer C, Grall A, Roussignol M (2003) Sequential condition-based maintenance scheduling for a deterioration system. Eur J Oper Res 150:451–461.

196. Castanier B, Grall A, Bérenguer C (2005) A condition-based maintenance policy with non-periodic inspections for a two-unit series system. Reliab Eng Syst Saf 87:109–120.

197. Nakagawa T, Yasui K (1991) Periodic-replacement models with threshold levels. IEEE Trans Reliab 40:395–397.

198. Ito K, Nakagawa T (1995) An optimal inspection policy for a storage system with three types of hazard rate functions. J Oper Res Soc Jpn 38:423–341.

199. Martines EC (1984) Storage reliability with periodic test. In:Proc Ann Reliab Maintainability Symp:181–185.

200. Gertsbakh I (2000) Reliability Theory with Applications to Preventive Maintenance. Springer, Berlin.

201. Dohi T, Kaio N, Osaki S (2003) Preventive maintenance models:Replacement, repair, ordering, and inspection. In:Pham H (ed) Handbook of Reliability Engineering. Springer, London:349–366.

202. Scarf PA (1997) On the application of mathematical models in maintenance. Eur J Oper Res 99:493–506.

203. Murthy DNP, Jack N (2003) Warranty and maintenance. In:Pham H (ed) Handbook of Reliability Engineering. Springer, London:305–316.

204. Adams PJ (1982) The quality of aircraft maintenance. Qual Assurance 8:87–95.

205. Rosenfield D (1976) Markovian deterioration with certain information. Oper Res 24:141–155.

206. Tijims H, Van der Duyn Schouten FA (1984) A Markov decision algorithm for optimal inspections and revisions in a maintenance system with partial information. Eur J Oper Res 21:245–253.

207. Hontelez JAM, Burger HH, Wijnmalen DJD (1996) Optimal condition-based maintenance policies for deteriorating systems with partial information. Reliab Eng Syst Saf 51:267–274.

208. Chen D, Trivedi KS (2005) Optimization for condition-based maintenance with semi-Markov decision process. Reliab Eng Syst Saf 90:25–29.

209. Saasouch B, Dieulle L, Grall A (2004) Maintenance policy of a system with several modes of deterioration:10th ISSAT Inter Conf Reliab Qual Design:211–215.

210. Liao H, Chan LY, Elsayed EA (2004) Maintenance of continuously monitored degrading systems:10th ISSAT Inter Conf Reliab Qual Design:216–220.

211. Elsayed EA, Zhang Z (2006) Optimum threshold level of degraded structures based on sensors data. In: 12 th ISSAT Inter Conf Reliab Qual Design:187–191.

212. Nakagawa T (1979) Optimal policies when preventive maintenance is imperfect. IEEE Trans Reliab R-28:331–332.

213. Nakagawa T (1979) Imperfect preventive-maintenance. IEEE Trans Reliab R-28:402.

214. Nakagawa T, Yasui K (1987) Optimum policies for a system with imperfect maintenance. IEEE Trans Reliab R-36:631–633.

215. Nakagawa T (2000) Imperfect preventive maintenance models. In:Ben-Daya M, Duffuaa SO, Raouf A (eds) Maintenance, Modeling and Optimization. Kluwer Academic, Boston:201–214.

216. Nakagawa T (2002) Imperfect preventive maintenance models. In:Osaki S (ed) Stochastic Models in Reliability and Maintenance. Springer, Berlin:125–143.

217. Wang H, Pham H (2003) Optimal imperfect maintenance models. In:Pham H (ed) Handbook of Reliability Engineering. Springer, London:397–414.

218. Nguyen DC, Murthy DNP (1981) Optimal preventive maintenance policies for repairable systems. Oper Res 29:1181–1194.

219. Nakagawa T (1986) Periodic and sequential preventive maintenance policies. J Appl Probab 23:536–542.

220. Nakagawa T (1988) Sequential imperfect preventive maintenance policies. IEEE Trans Reliab 37:295–298.

221. Mie J (1995) Bathtub failure rate and upside-down bathtub mean residual life. IEEE Trans Reliab 44:388–391.

222. Baker Jr HG (1978) List processing in real time on a serial computer. Commun ACM 21:280–294.

223. Steele Jr GL (1975) Multiprocessing compactifying garbage collection. Commun ACM:18:495–508.

224. Cohen J (1981) Garbage collection of linked data structures. ACM Comput Surv 13:341–367

225. Kung HT, Song SW (1977) An efficient parallel garbage collection system and its correctness proof. In:18th Ann IEEE Symp Found Comput Sci:120–131.

226. Lieberman H, Hewitt C (1983) A real-time garbage collection based on the lifetimes of objects. Commun ACM 26:419–429.

227. Yuasa T (1990) Real-time garbage collection on general-purpose machines. J Syst Software 11:181–198

228. Suzuki K, Nakajima K (1995) Storage management software. Fujitsu 46:389–397.

229. Velpuri R, Adkoli A (1998) Oracle 8 Backup and Recovery Handbook. McGraw–Hill, England.

230. Chandy KM, Browne JC, Dissly CW, Uhrig WR (1975) Analytic models for rollback and recovery strategies in database systems. IEEE Trans Software Eng SE-1:100–110.

231. Reuter A (1984) Performance analysis of recovery techniques, ACM Trans Database Syst 4:526–559.

232. Young JW (1974) A first order approximation to the optimum checkpoint interval. Commun ACM 17:530–531.

233. Gelenbe E (1979) On the optimum checkpoint interval. J Assoc Comput Machinary 26:259–270.

234. Fukumoto S, Kaio N, Osaki S (1992) A study of checkpoint generations for a database recovery mechanism. Comput Math Appl 1:63–68.

235. Nakagawa S, Fukumoto S, Ishii N (2003) Optimal checkpointing intervals of three error detection schemes by a double modular redundancy. Math Comput Model 38:1357–1363.

236. Qian CH, Pan Y, Nakagawa (2002) Optimal policies for a database system with two backup schemes. RAIRO Oper Res 36:227–235.

237. Nakamura S, Qian CH, Fukumoto S, Nakagawa T (2003) Optimal backup policy for a database system with incremental and full backups. Math Comput Modelling 38:1373–1379.

238. Parzen E (1962) Stochastic Processes. Holden-Day, San Francisco.

239. Beekman JA (1974) Two Stochastic Processes. Almqvist and Wiksell International, Stochholm.

240. Becker N (1989) Reliability models in cancer epidemiology. Biom J 31:727–748.

241. Becker N, Rittgen W (1990) Some mathematical properties of cumulative models regarding their application in cancer epidemiology Biom J 32:3–15.

242. Smith W (1973) Shot noise generated by a semi-Markov process. J Appl Probab 10:685–690.

243. Rice J (1977) On generalized shot noise. Adv Appl Probab 9:553–565.

244. Hsing TL, Teugels JL (1989) Extremal properties of shot noise processes. Adv Appl Probab 21:513–525.

245. Doney RA, O'Brien GL (1991) Loud shot noises. Ann Appl Probab 1:88–103.

246. McCormik WP (1997) Extremes for shot noise processes with heavy tailed amplitudes. J Appl Probab 34:643–656.

247. Lund RB, Butler RW, Paige RL (1999) Prediction of shot noise. J Appl Probab 36:374–388.

248. Lund RB, McCormik WP, Xiao YH (2004) Limiting properties of Poisson shot noise processes. J Appl Probab 41:911–918.

249. Waymire E, Gupta VK (1981) The mathematical structure of rainfall representations 1:A review of the stochastic rainfall models. Water Resourc Res 17:1261–1272.

250. Moran PA (1967) Dams in series with continuous release. J Appl Probab 4:380–388.

251. Yeh L, Hua LJ (1987) Optimal control of a finite dam. J Appl Probab 24:196–199.

252. Abdel-Hameed M, Nakhi (1990) Optimal control of finite dam using $P^M_{\lambda,\tau}$ policies and penalty cost: Total discounted and long run average cases. J Appl Probab 28:888–898.

253. Lund RB (1994) A dam with seasonal input. J Appl Probab 31:526–541.

254. Harrison JM, Resnick SI (1976) The stationary distribution and first exit probabilities of a storage process with general release rule. Math Oper Res 1:347–358.

255. Prabhu NU (1980) Stochastic Storage Processes. Springer, New York.

256. Lund RB (1996) The stability of storage models with shot noise input. J Appl Probab 33:830–839.

257. Lemoine AJ, Wenocur ML (1986) A note on shot-noise and reliability modeling. Oper Res 34:320–323.
258. Rolski T, Schmidli H, Schmidt V, Teugels J (1999) Stochastic Processes for Insurance and Finance. John Wiley & Sons, Chichester, England.
259. Klüppelberg C, Kyprianou AE, Maller RA (2004) Ruin probabilities and over-shoots for general Lévy insurance risk processes. Ann Appl Probab 14:1766–1801.
260. Embrecht P, Klüppelberg C, Mikosch T (1997) Modelling Extremal Events for Insurance and Finance. Springer, Berlin.
261. Asmussen S (2001) Ruin Probabilities. World Scientific, Singapore.
262. Williams T, Ancker Jr CJ (1963) Stochastic Duels. Oper Res 11:803–817.
263. Bhashyam N (1970) Stochastic duels with lethal dose. Nav Res Logist Q 17:397–405.
264. Bhashyam N (1973) Stochastic duels with correlated fire. Metrika 20:17–24.
265. Thompson DE (1972) Stochastic duels involving reliability. Nav Res Logist Q 19:145–148.
266. Lucas TW (2003) Damage functions and estimates of fratricide and collateral damage. Nav Res Logist 50:306–321.
267. Manor G, Kress M (1997) Optimality of the greedy shooting strategy in the presense of incomplete damage information. Nav Res Logist 44:613–622.

Index